Beachcombing

A guide to seashores of the Southern Hemisphere

Ceridwen Fraser

OTAGO UNIVERSITY PRESS
Te Whare Tā o Te Wānanga o Ōtākou

Note for teachers: Teacher notes are available at:
https://www.otago.ac.nz/press/books/otago824383.html

Published by Otago University Press
533 Castle Street
Dunedin, New Zealand
university.press@otago.ac.nz
www.otago.ac.nz/press

First published 2021

ISBN 978-1-99-004800-5

A catalogue record for this book is available from the National Library of New Zealand.

Editor: Imogen Coxhead
Index: Imogen Coxhead
Design and typesetting: Fiona Moffat

Artwork and photos in this book were created by the author unless otherwise indicated.

Front cover photo: Ceridwen Fraser
Author photo: Alan Dove

Printed in China through Asia Pacific Offset

CONTENTS

A gentoo penguin waddles out of the sea near some beach-cast southern bull kelp, *Durvillaea antarctica*, in the Falkland Islands. Southern bull kelp can drift tens of thousands of kilometres at sea, carrying diverse animals across entire ocean basins.

Introduction

Beaches are our windows to the ocean. We walk along beaches and gaze at the heaving, shimmering waves, wondering about what lies beneath. The ocean changes from day to day – from crystal clear and calm one day to dark, murky and violent the next. Beaches, too, change rapidly – sand dunes shift and pebbles and sometimes huge boulders appear and disappear. Ancient shipwrecks are exposed and then reburied, almost in the blink of an eye. As sands move and tides recede, clues to the life-and-death dynamics of the marine realm are cast up on the beach – random objects that we poke and puzzle over. Each of these objects tells a story.

The coasts of the Southern Hemisphere are connected by strong ocean currents. There are many similarities, as well as some differences, in the species that live around these coasts. This book is a general guide to beachcombing in the Southern Hemisphere – in particular, but not exclusively, the relatively high latitude temperate and cold-temperate regions. Although many of the examples are from Australia and New Zealand, they are relevant too for Africa, South America and Southern Hemisphere oceanic islands.

After highlighting beach conservation concerns and Indigenous rights, the book gives an overview of near-shore ocean dynamics – the tides, waves, winds and ocean currents that push objects around and can cause them to wash up on beaches. Then, a mention of some of the high-value treasures you might stumble across during beach wanderings, such as ambergris and gold. Much of the rest of the book deals with the diversity of ocean life – with a focus on the plants and animals that turn up or live on and around sandy beaches and the rock platforms that flank them, and what they tell us about life and death in the sea. Finally, a look at connections between Southern Hemisphere coasts, and the future of southern seashores.

This book is for browsing – for picking up and putting down, and poking into or flicking through when you want to know more about things you have found on the beach. Suggestions for further reading on particular topics, and for identification guides, are provided in the last section.

Some beach-dwelling birds, such as oystercatchers, make their nest on the sand. Their eggs can be hard to see.

Looking After Our Beaches

What you find on a beach is not necessarily yours to take home. Beach ecosystems are complex, and empty shells, washed-up seaweed and other objects have important roles to play in nutrient cycles and sedimentation. Collecting a few dead and uninhabited shells or other non-living objects from the beach is generally okay, but in some places the removal of anything is strictly prohibited – especially in marine protected areas. In New South Wales, Australia, for example, in national parks, nature reserves and Aboriginal areas collecting shells is prohibited; in most other parts of NSW you can collect small quantities of shells and shell grit. Protected areas are usually well signposted, but check online for detailed information about what is allowed for the beaches you are visiting.

Regardless of whether or not an area is protected by regulations, when visiting beaches it is important to be aware of the impact you might be having on local wildlife. As tempting as it can be to try to get up close and personal with charismatic species such as penguins and seals, your approach will cause them stress. Research shows that a penguin's heart rate increases dramatically when humans are in view, causing them to use up more energy. Young may also become separated from their parents, increasing stress levels and the risk of harm.

As a beachgoer you also need to watch your step. Some birds lay their eggs in the sand, and these are usually hard to see and easy to destroy. On rock platforms near beaches, too, your footsteps may damage seaweeds and marine invertebrates.

Coastal ecosystems provided, and continue to provide, diverse resources for Indigenous inhabitants, and beachgoers must consider and respect the significance of coastal areas to Indigenous people. In Australia, coastal areas ('sea country') were densely populated before European colonisation; likewise, the Māori in New Zealand, and the Lafkenche and Williche (coastal Mapuche) in Chile, have a strong cultural connection to the ocean and coastal areas: both, for example, have oral legends of people or spirits riding on whales. Historically, Indigenous people often used the same places year after year for gathering and feasting on seafood, coastal plants and animals. The hard remains of these activities, such as mollusc shells and bird and seal bones, often last for centuries in middens. The

The painting *Guiding Spirits* by Alvin Pankhurst, 2017, shows sculptures of Māori spirits on a beach.

layers of middens provide insight into what ecosystems were like in the past, and how food resources have varied through time. We usually can't see back more than a few hundred years in coastal middens, though; the melting of ice caps after the last ice age (which reached its coldest point around 20,000 years ago) has since led to sea-level rise of, on average, about 120m, obscuring many ancient coastlines.

This book uses mainly English-language terms for the things it describes, but across the Southern Hemisphere there are hundreds of Indigenous languages. The table opposite offers just a few of the various words for beaches and the animals, plants and objects found on them, but these necessarily represent only a tiny handful of examples from the southern Pacific region. Within Australia alone there are hundreds of Aboriginal languages; this table shows a few words from just two. One is palawa kani, the language of Tasmanian Aboriginal people, and the other is Wonnarua, a language from the Hunter Valley region of New South Wales. Also shown are some words from te reo Māori (from Aotearoa New Zealand) and Mapudungun from Chilean Mapuche Lafkenche.

Examples of beach-related words from some Southern Hemisphere Indigenous languages. This table only hints at the enormous diversity of terms that exist. (With thanks to the Tasmanian Aboriginal Centre, the Wonnarua Nation Aboriginal Corporation, the Lafkenche lonko and www.maoridictionary.co.nz.)

	AUSTRALIA: HUNTER REGION	AUSTRALIA: TASMANIA	NEW ZEALAND	CHILE
ENGLISH	WONNARUA	PALAWA KANI	TE REO MĀORI	MAPUDUNGUN
fish	makurr	pinungana	ngohi	chajwa/pekan
fishing net	tarila		kupenga	ñeweñ
ocean	kar(u)wa	muka	moana	l̲afken̲
waves		nipakawa	ngaru	rew
rough sea	kar(u)wa tjalkan		karekare	aguley tilafken
sand	puna	nguna	onepū	küyim
beach	wampal	katina	tāhuna	inalafken
wood/driftwood	pulpi	mumara	paewai/porotāwhao	mamüll ina lafken
abalone		nitipa	pāua	mañiü
sea urchin		luwanuma	kina	yüpe
sea star		wiri		wanglen
seaweed	kayarr	ruri	rimurimu	
bull kelp, *Durvillaea*		rikawa	rimurapa	coyofe/kollof
mussel		miri	kuku	chorro
crab	tinting	tuti	pāpaka	koinau
cuttlefish	kar(u)wa tjalan	wimini		
oyster	munpunkan	taralangkana	tio repe/tio para	pellü
seagull	ngawu	ruwinana	tarāpunga	chillwe/kaleu
sea lion			rāpoka/whakahao	lamme
shark	karakayung	wayana	mangō	
shell	kaling	rina	kota	chakantü
sperm whale/whale	payinapakan	maytawinya	pāmu wēra/parāoa	yene
ambergris			mīmiha/ tūtae tohoraha	mélame kadun

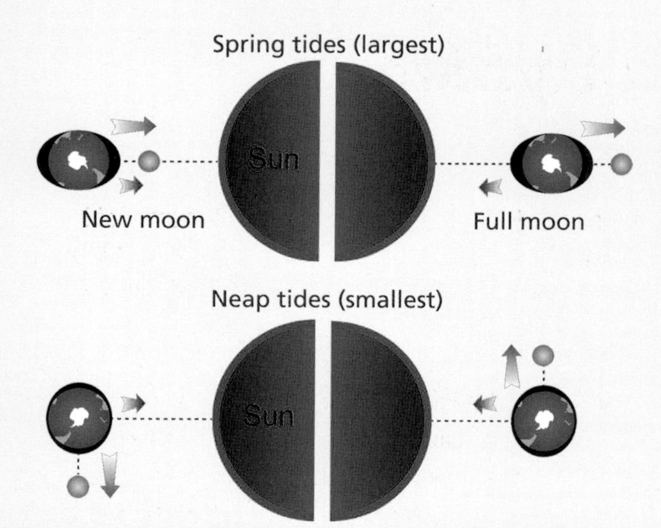

LEFT: Tides result from the combined gravitational pull of the moon (mostly: grey arrows) and the sun (to a lesser extent: orange arrows) on the Earth's oceans. The largest (spring) tides occur when the sun is in line with the moon and the Earth.

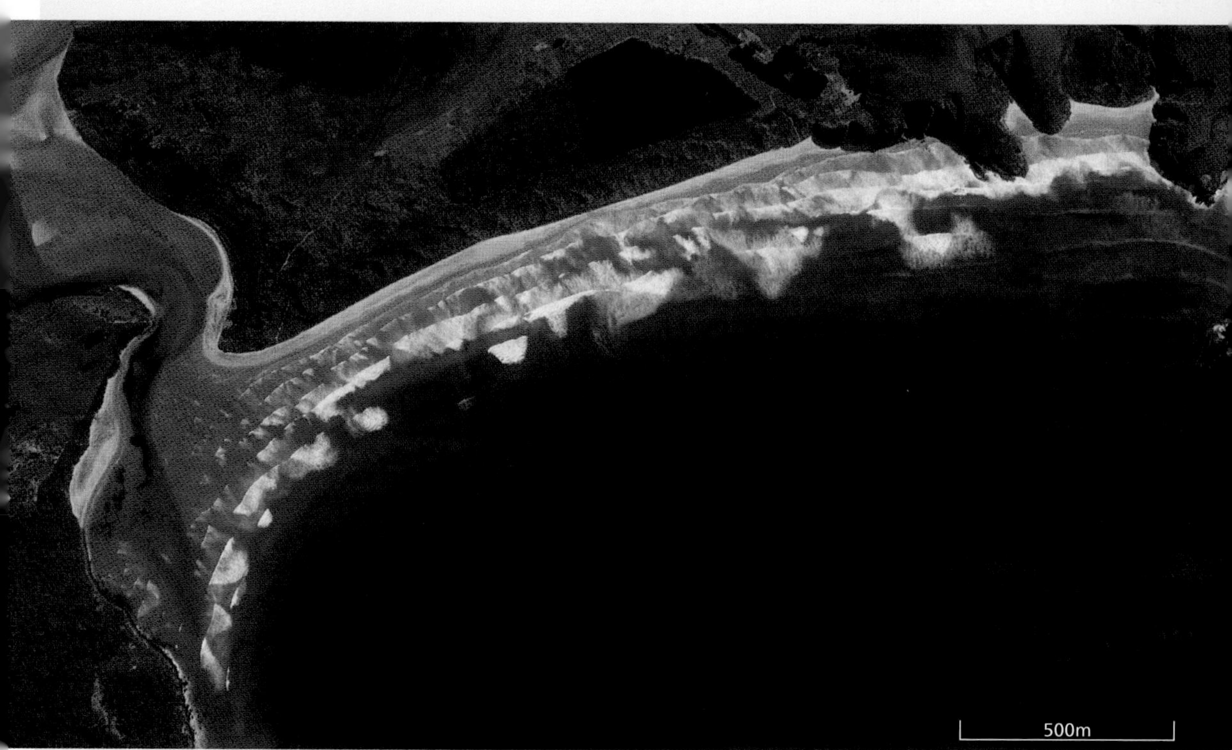

Waves refracting (bending) to align with a beach in New Zealand. *Image: Google Earth, November 2020*

Coastal Ocean Dynamics

TIDES

Arguably one of the best times to find interesting things washed up on a beach is as the tide is going out.

The moon might look small in the sky, but its diameter is roughly one quarter and its mass about one percent that of Earth. Because it's so close to Earth, the gravitational pull of its mass has a strong impact on our planet. Much of the surface of the Earth is covered by oceans, and these large bodies of water bulge toward the moon on the moonward side, and away from the moon on the other side. As the Earth spins, different parts of the world's oceans bulge. When your part of the ocean is in a bulge, the water will be high along the coast (high tide), and when you are furthest from the bulge, you experience low tide. Because the bulge is on both sides of the Earth, and the Earth takes a day to spin on its axis, most beaches will experience two high tides and two low tides every day. There are some exceptions, where the shape of the land interferes with the twice-daily rise and fall of the water.

The sun also has a gravitational pull on the Earth, albeit a relatively small one because it is so much further away. When the Earth, moon and sun line up, the forces work together and the tides are particularly big – very high and very low (spring tides). When the moon, Earth and sun are at right angles, the solar and lunar gravitational forces act against each other and tides are small (neap tides). Because Earth's orbit around the sun is not circular, spring and neap tides vary in intensity, with the biggest spring tides occurring at

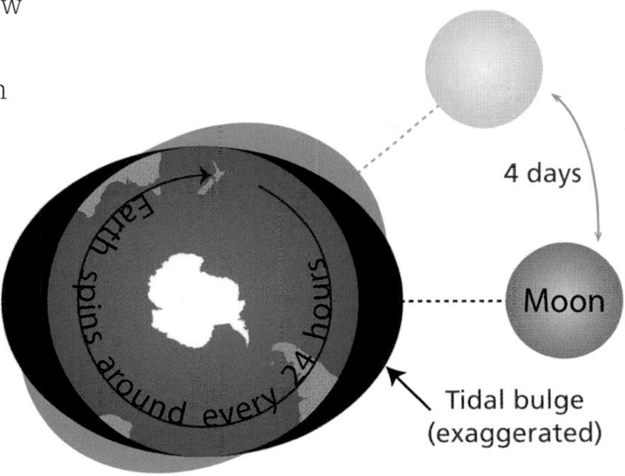

RIGHT: Tide times change each day as the moon moves through its month-long orbit of the Earth. Water bulges toward the moon and – on the opposite side of the Earth – away from the moon. Where the bulge is greatest the tide will be high.

the beginning of the year (king tides). How much the water rises and falls around the world varies greatly, as the flow of water is not only affected by the moon (and the sun) but also by the shapes of the continents and the depth of the ocean. Water heights are also affected by the tilt of the Earth and the weather.

The moon orbits the Earth every 27.3 days. If you visit a beach at the same time each day, the moon's different position in orbit (and to a lesser extent, the sun's influence) means that the high and low tide times will have changed. Generally, tides occur about one hour later each day – so if you visit a beach at midday one day and notice the tide is high, you can expect it to be high at around 1pm on the following day. To get an accurate estimate of tide times and heights, you can look online – government science and weather services provide useful online tide calculators that can tell you almost exactly when tides will occur at a given beach, each day, and how high or low they will be.

WAVES

Wind causes most of the ocean waves that affect the waters near the coast. Surface waves can travel a long way, and the wind that started the breakers that are smashing on your local beach might have been blowing hundreds or even thousands of kilometres away. By the time waves have come so far they have generally settled into a fairly regular height (the vertical measure from crest to trough, which is double the amplitude – the measure from crest to rest) and period (the horizontal distance between waves). Strong winds close to shore can, however, create a jumble of waves of different heights and periods, which can make coastal waters unpredictable and rough – not great for surfing!

As a wave nears the shore the decreasing water depth makes it slow down, causing wave energy to pile up and increasing wave height. Water molecules inside a wave have a circular movement when out at sea, but closer to shore the friction from the seabed causes the base of a wave to move more slowly. The top part moves faster than the bottom until the tall crest reaches over the trough, eventually falling and making the wave 'break'.

Because a wave slows down as water becomes shallow, the parts of a wave near the beach travel more slowly than parts of the same wave in deeper, more offshore water. The slowing section then drags part of the waveline around to be parallel to the beach. This change of direction, or refraction, helps to explain why breaking waves generally line up with a beach even if the swell direction further out is not parallel with the shore.

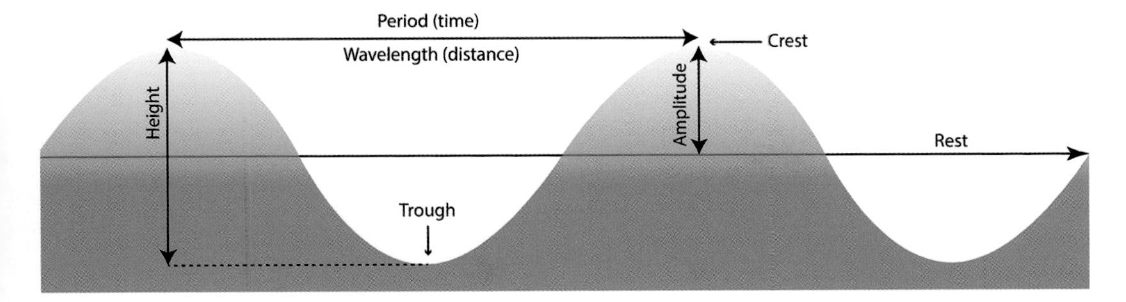

The characteristics of waves.

Sometimes waves from different directions interact and the combined energy can cause interference patterns. As a result some waves might be much bigger (constructive interference) or smaller (destructive interference) than others in a set.

Small waves tend to help deposit sand on beaches, whereas large waves can erode beaches – sometimes dramatically. A powerful storm might alter the shape of a beach by removing sand, and can even turn a sandy beach into a pebbly or bouldery beach almost overnight. Waves also have a strong influence on what washes up on the shore, and storms may lead to mass strandings of marine organisms and debris from far out at sea.

Waves can also make beaches dangerous: large high-energy waves pose a risk to even the strongest of swimmers. It's always good to keep half an eye on the ocean while beachcombing.

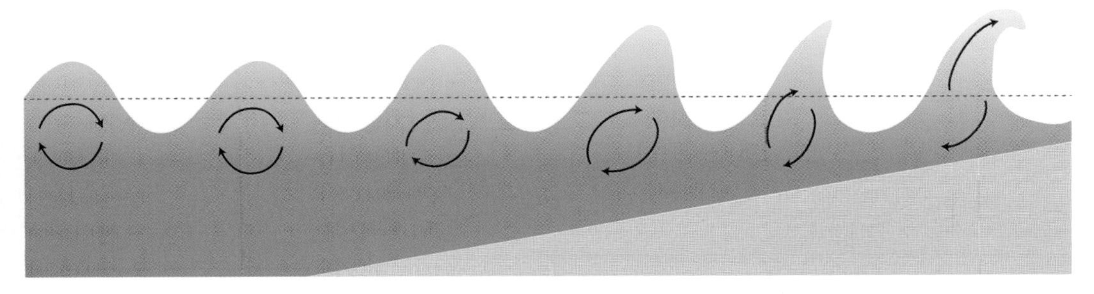

Waves break near the shore because the bottom of the wave is slowed by friction on the seabed, pushing energy upwards.

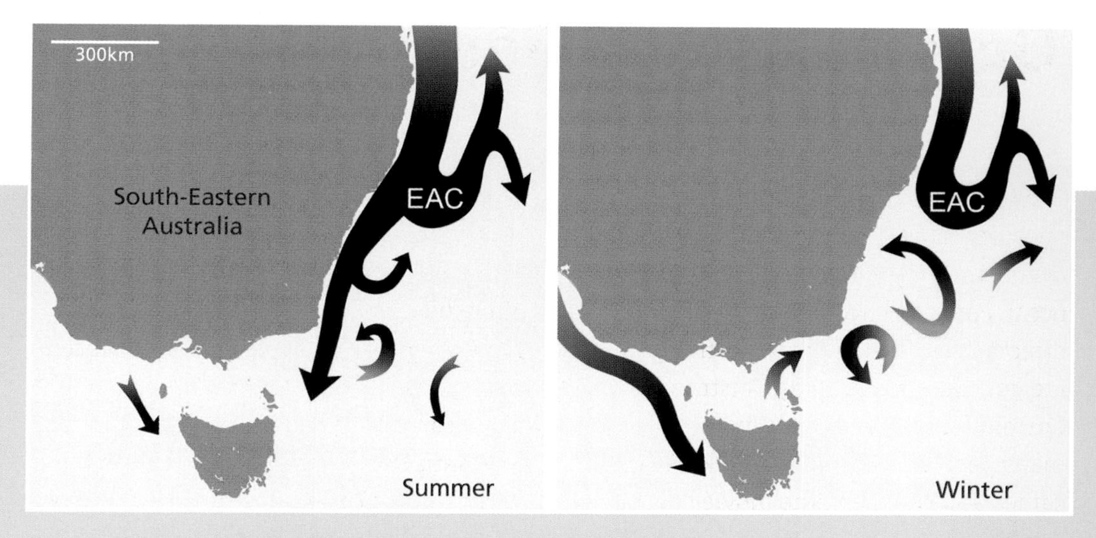

Ocean currents are dynamic and can change direction and strength. In summer the East Australian Current extends further south than in winter, when its southward flow is weaker and is disrupted by eddies.

For hundreds of years people have used the ocean to carry messages in bottles.

Ocean Current Oddities

The things we find on a beach can tell us a great deal about ocean dynamics. Oceans are ever-changing, and although major ocean currents have largely predictable directions and strengths, even the largest are not constant. The East Australian Current (EAC) for example (with which many people are familiar, thanks to its cameo appearance in the movie *Finding Nemo*), is a strong current running southward along the eastern coast of Australia. In winter, however, the EAC is often disrupted by eddies in the southern parts of its range that can lead to localised northward, rather than southward, dispersal of larvae and flotsam. In the Southern Ocean, the eastward flow of the Antarctic Circumpolar Current is disrupted

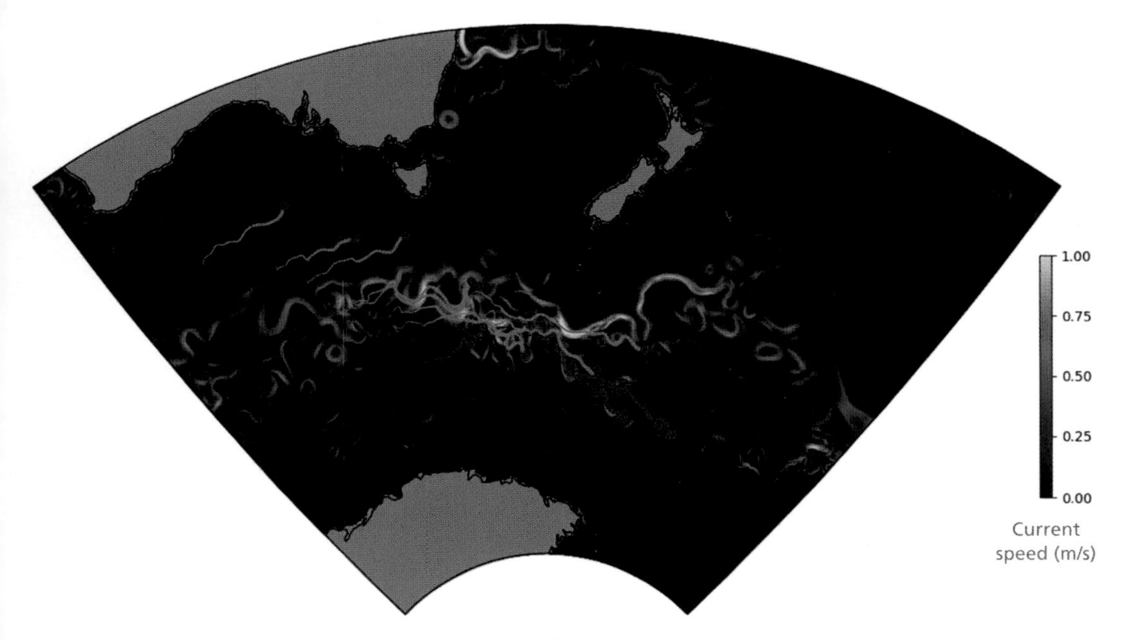

The simulated trajectories of floating particles released from South Georgia show how unpredictable oceanic dispersal can be. The particles took many different pathways and some ended up thousands of kilometres apart. In this simulation, those with green tails eventually reached the Antarctic coast (see Further Reading: Fraser et al. 2018). *Image: Adele Morrison and Andy Hogg*

at times by large eddies that can push packets of water (which might contain larvae and other animals) to the north or south of circumpolar oceanic fronts. These eddies are thought to play important roles in the movement of some species into or out of Antarctic waters.

Ocean water movement is affected to varying degrees by wind, temperature and a range of other factors, making it nearly impossible to predict where something that is cast adrift will end up. High-resolution particle dispersion modelling can be helpful for understanding the most likely general trajectories for drift particles, but there is a huge amount of possible variation.

MESSAGES IN BOTTLES

The romantic idea of putting a message in a bottle and throwing it into the ocean for someone to find draws directly on the unpredictability of surface currents. Where will the bottle end up, and who will ultimately read the message?

On 1 January 1916 three messages in separate bottles washed up on one beach in Victoria in southeast Australia, and another on a nearby beach. All had been thrown into the ocean by soldiers on a ship departing for Europe and the First World War. One was from a young soldier:

At sea, Saturday, December 25, 1915, 4pm

My dear Mum,

I am sending this note by bottle from the Victorian coast. I hope you will get this OK. We have just finished our Christmas dinner – turkey and pork. Everyone on board is OK. A girl was found on board dressed as a soldier; she was going to fight with her brother at Gallipoli. Oh, well, goodbye for the present.

I am your loving son,
Ted.

The woman discovered on the ship was teenager Maud Butler. In an interview with a local newspaper, the *Bendigo Independent* (29 December 1915), she said, 'It is not correct that I joined the ship just in sport, to see my brother ... My object was to do what I could to help. I wanted to join the Red Cross ... when I failed, I bought a khaki suit and stowed away.' According to the article, to avoid the sentry at the ship's gangway she climbed a rope that hung from the bow. Her undoing lay in the fact that she had only been able to get black boots instead of the regulation tan, and this anomaly brought her to the notice of an officer, who asked to see her identification. She was sent back to shore on a passenger ship.

On the western side of Australia in April of the same year, four soldiers threw a bottle into the sea with a message describing the great time they were having

Sealed bottles float at the surface and follow ocean currents.

as they set sail for Egypt. The bottle was found on a beach near Albany in Western Australia in the 1930s.

Another message in a bottle was found on a beach in Western Australia 132 years after it was set adrift from a German naval ship. This message was part of an early German scientific effort to understand how ocean currents move floating material around. Thousands of bottles had been released containing messages that gave the ship's coordinates and asking the finders to send details of where the bottles ended up.

Parts of a shipwreck that wash ashore are called flotsam.

OPPOSITE: A lifebuoy from the SS *Lurgurena*, found on a beach in Australia.

FLOTSAM AND JETSAM

A message in a bottle is a form of jetsam – something that is deliberately thrown overboard. In the past, a ship's crew would sometimes jettison heavy cargo in order to reduce the risk or speed of sinking – for example if the ship's hull was breached. Flotsam, in contrast, refers to floating material from a wreck, which was not intentionally jettisoned. Sometimes flotsam appears on beaches long after a ship is wrecked, as the sunken vessel slowly degrades and breaks apart, releasing buoyant parts.

Flotsam can tell us a great deal about ocean currents. When Malaysia Airlines flight MH370 tragically went missing without trace in 2014, a global search led to a huge investment in oceanographic research. In July 2015 a flaperon – a control flap from a plane's wing – was found covered in large goose barnacles on a beach in La Réunion. Biochemical analyses of the goose barnacles gave clues about the temperature of the water the flaperon had journeyed through, and showed it had spent some time in cool-temperate waters (18–20°C) before moving back into warmer (~25°C) waters. Other pieces of the plane were subsequently found on beaches along the eastern coasts of Africa. The locations of the debris and analysis of Indian Ocean current dynamics allowed scientists to determine the area where the plane must have crashed. The enormous international efforts that went into trying to locate the crash site have greatly improved our knowledge of ocean dynamics and dispersal of drift material – as well as improving our maps of the Indian Ocean sea floor.

In another example, 28,000 rubber duckie bath toys fell off a cargo ship in the middle of the North Pacific in 1992. These have since washed up on beaches all around the world, including in the Atlantic and South Pacific oceans. Information on where they have been found has helped us to understand oceanographic patterns and processes, including how long floating objects caught up in large gyres can continue to circle oceans.

Flotsam and jetsam generally sit in the 'finders keepers' basket under maritime law, unless someone is able to demonstrate that they are the rightful owner. Most of the buoyant parts of shipwrecks – such as lifebuoys – are not particularly valuable, but each has a story to tell, and some hint at human tragedies and sunken treasure lost in the ocean depths.

Part of a space rocket found on a beach in southern New Zealand in 2004. The piece is made of titanium, weighs 30kg, and is hollow and buoyant. *Photo Lloyd Esler*

GOOSE BARNACLES

Almost anything that has spent at least several weeks drifting near the surface of the ocean will have **goose barnacles** attached. Barnacles are often mistaken for molluscs (the group that includes snails and mussels) because of their hard external shells, but they are actually crustaceans (related to crabs, lobsters and krill) that as adults are sedentary – in other words, they stick to one place. They have cirri, appendages like legs that resemble a fan or a rake, which they sweep through the water to collect particles of food.

Goose barnacles are 'stalked' barnacles – their shells make a flat, teardrop-shaped compartment at the end of a fleshy stalk. Some goose barnacles grow on rocks on the edge of the shore, but those from the genus *Lepas* usually only grow on pelagic objects – things drifting out at sea. The larvae of *Lepas* settle on almost any drifting material, including driftwood, kelp and plastics.

Because *Lepas* goose barnacles do not normally attach to objects on or close to the shore, their size can be a useful indicator of the minimum time that an item has spent drifting in the open ocean. *Lepas* growth rates have, for example, been used to estimate how long drifting kelp took to travel between sub-Antarctic Campbell Island and the South Island of New Zealand: the journey had taken more than two months.

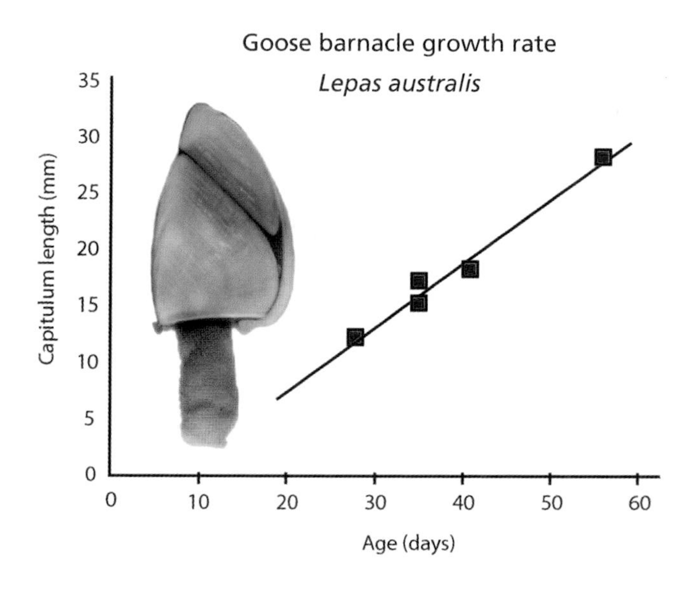

Goose barnacles (*Lepas*) are crustaceans that grow on materials drifting at sea. Their size helps to show how long the object has been at sea. This graph, from a study by Skerman in 1958 (see Further Reading), shows the growth rate of *Lepas australis* goose barnacles, based on capitulum length (from the tip to the base of the shelled part).

ABOVE: Goose barnacles on bull kelp that had been drifting at sea before washing up on a beach.

LEFT: Goose barnacles on a piece of pumice found on a beach in Australia. Note the extended foot collecting food from the water. *Photo John Walker*

PLASTICS

Plastic debris often washes up on beaches – a tiny tip-of-the-iceberg indication of the amount of plastic in the oceans. A great deal of it is single-use packaging from food and drink – wrappers, bottles and plastic bags – and cigarette butts. The term 'plastic' really means having the ability to be moulded from one shape to another, but we commonly use the term to refer more specifically to organic polymers, usually but not always derived from petroleum and with other compounds added to achieve particular strength or malleability. Much single-use plastic will take tens or hundreds of years to decompose, so the accumulation of plastics in the ocean is a problem not easily solved by time alone, especially as the amount of plastic entering the ocean is estimated at several million tonnes per year.

Fragments smaller than 5mm are termed 'microplastics', are easily ingested by animals and can release toxins into the food chain. These toxins are bad for the health of marine life and, in some cases, make their way onto our dinner tables too. Primary microplastics, such as the microbeads in cosmetics or fibres from synthetic clothing, are small to start with. Secondary microplastics are the broken-up pieces of large plastic items, such as parts of a plastic bag or other packaging.

Plastics and other floating materials can accumulate in windrows (lines parallel to the wind) at the surface of the ocean, as the wind creates linear convergence zones via a process called 'Langmuir circulation'.

Plastic rubbish on a beach in Tahiti. *Photo Chris Woods*

Large plastics can also directly harm ocean life. Discarded plastic fishing nets (ghost nets) often entangle fish, turtles and marine birds and mammals. Soft plastics such as shopping bags sometimes resemble sea jellies or other prey and are swallowed by large animals. Autopsies of beached marine creatures such as whales, dolphins and sea turtles often reveal large amounts of plastic debris in their guts.

A synthetic sponge masquerading as the real thing on a beach in New South Wales, Australia.

Plastics are found throughout the ocean and on beaches around the world, including in Antarctica. *Photo Kerryn Wood*

Some plastics are heavier than seawater and sink to the ocean floor, where their greatest impact is on bottom-dwelling ('benthic') animals. Others float, and these floating plastics are driven by ocean currents and may accumulate in clumps known as 'plastic islands' or 'garbage patches'. Most ocean basins have garbage patches in the calm centre of their oceanic gyres; perhaps the best known is the Great Pacific Garbage Patch (actually two patches, one to the east and one to the west) in the North Pacific Ocean.

Onshore winds can lead to sudden dumps of a lot of plastic debris onto beaches. Blaming others for the rubbish in the ocean is easy, but plastics often have telltale signatures that show where they have come from, such as labels showing which country they were sold in, and close inspection often reveals that a good deal of the plastic on our beaches has come from our own country. Cigarette butts, food packaging and other waste discarded on the street in coastal cities can be washed down drains or blown directly into the ocean.

We all have a responsibility to reduce the amount of plastic making its way into the sea, and the best way to do this is to use less plastic.

WHERE THE WIND BLOWS

Floating hydroids with stinging tentacles, known as **bluebottles** or **Portuguese men of war** (*Physalia physalis*), or **by-the-wind sailors** (*Velella velella*), drift at the surface of the ocean and are often blown on shore, sometimes in mass strandings.

The sting of the bluebottle can be painful and can cause an allergy-like reaction in some people. If there are many fresh bluebottles on the beach it's probably best not to go swimming, since more are likely to be washing in. You should also watch where you step, as the tentacles can still sting while wet on the beach. Dried bluebottles are generally okay to touch, and you can handle a living bluebottle safely by holding its sail, the gas-filled bladder, which has no stinging cells. *Velella* can also sting humans, but the effect is comparatively mild.

Single shoes are often found on the beach, perhaps because the shape of left and right shoes sends them in different directions at sea.

Photo Chris Woods

Each apparent individual is actually a colony made up of several individuals that have specialised roles in reproduction, feeding and defence. Some form tentacles that dangle down in the water and catch passing fish with venomous stinging cells, or nematocysts, which immobilise prey. The tentacles then retract, pulling the prey up to the feeding polyps (other individuals in the colony), which digest them.

At the top of the colony, above the cluster of reproductive and feeding polyps, is the sail. In *Physalia* the sail is shaped a little like the bag of a set of bagpipes and tapers to a point at either end. In some colonies the sail may be so large that it extends up to 15cm above the surface of the water. In *Velella*, the sail is a rigid sheet of transparent tissue rather than a gas-filled sac. In both *Physalia* and *Velella*, the sails have a slight twist to them. Interestingly, some twist to the left and others to the right.

ABOVE: Stinging bluebottles often turn up on beaches. *Photo Chris Woods*

LEFT: Bluebottles (*Physalia*) are colonies of hydroids that float at the ocean surface, and use their tentacles to catch prey.

Several hypotheses have been put forward to explain why there is this left/right difference in sails. There's no doubt, however, that the twist affects where the animals go: both *Physalia* and *Velella* depend almost entirely on the wind to move, but some move to the left of the downstream wind direction while others move to the right, meaning colonies head off in different directions when the wind blows. This dispersal probably helps them to maximise access to food resources so they're not all feeding in the same place, and would also help to safeguard against having them all wash up and die on the same beach at the same time. The influence of wind on left/right dimorphism in floating objects could also explain why you rarely find two matching shoes among the flotsam on a beach!

One beautiful predator of pelagic hydroids such as *Physalia* and *Velella* is *Glaucus*, a blue sea slug (nudibranch) that can sometimes wash up on a beach, especially during mass strandings of its prey. There are several species in the genus, but they all look similar and have similar lifestyles.

Glaucus is a nudibranch that preys on bluebottles and other drifting hydroids.

Glaucus has an internal air sac that helps it to float at the surface, technically upside down, with its true dorsal (back) side facing down into the water and its ventral (tummy) side facing the sky. The ventral side is a stunning patterned blue with light and dark stripes, and this colouring helps *Glaucus* to blend in with the water surface when viewed from above, protecting it from predators such as birds. The dorsal side is a silvery grey, which blends in with the sky when viewed from below. *Glaucus* is immune to the venomous *Physalia* or *Velella* nematocysts, and in fact can incorporate these stinging cells into its own tissue as a defence mechanism. That means that you should be wary of touching *Glaucus* with bare hands, as it can sting. The nematocysts that *Glaucus* gathers from its prey are held at the tips of its long, finger-like structures (cerata), which the animal also uses for swimming.

CRUSTACEAN TIDES

Crustaceans are animals with several limbs and a hard external coating (exoskeleton) around their bodies; the group includes shrimps, crabs and lobsters. They're arthropods, distantly related to land-based arthropod groups such as insects – slaters (woodlice) are actually land-based crustaceans. Sometimes you might notice bands of thousands of small dead or

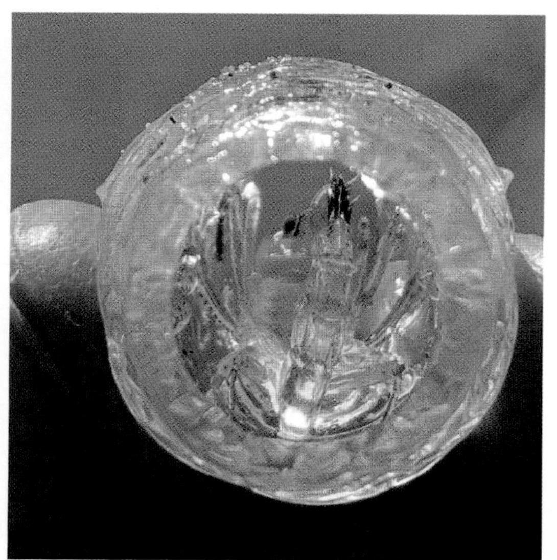

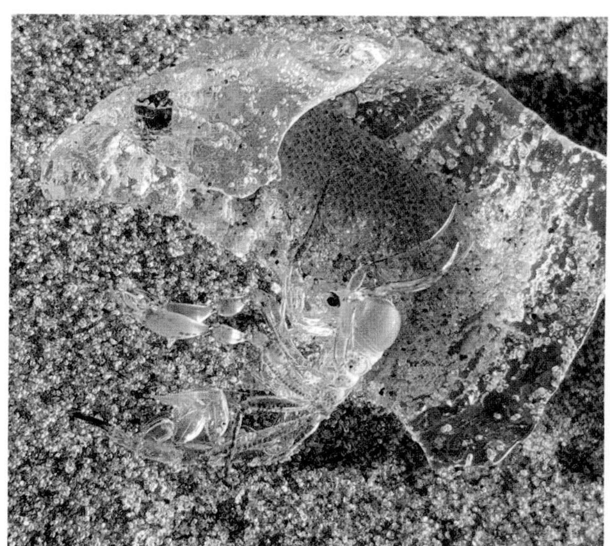

Phronima is a striking hyperiid amphipod that parasitises salps and resembles the queen in the movie *Alien*. You can sometimes find these parasitoid amphipods, and their eggs, inside the barrel-like remains of salps washed up on the beach.

dying crustaceans washed up at the edge of the waves on a beach. These could be **hyperiid amphipods**, some of which are characterised by large eyes that help them see in mid-ocean depths. Others are nearly invisible, with transparent bodies and pigment only in their eyes. Some have even developed sophisticated anti-reflective layers on their external cuticle – tiny hair-like protuberances or subwavelength-sized spheres – to enhance their invisibility. These species are pelagic, living in the open ocean, sometimes quite deep, and many are parasites of gelatinous animals such as salps or jellyfish.

Hyperiid amphipods are particularly abundant at colder, higher latitudes (in the Southern Hemisphere this means more southern areas). In the northern parts of the Southern Ocean they can be so abundant that they become one of the major food sources for squid, fish, marine mammals and seabirds. They are not particularly strong swimmers, and can be moved around by ocean currents and mesoscale eddies. When a body of water moves rapidly from the mid-ocean depths to the shallows – an upwelling – large numbers of plankton, including hyperiid amphipods, can be brought suddenly to the near-surface layers. Their exoskeletons are thought to make them highly buoyant at the air–water interface, essentially trapping them at the surface and leading to their being washed ashore in droves.

Hyperiid amphipods are not the only swarming or schooling crustaceans that get washed up. In New Zealand and southeastern Australia, both **white krill** (*Nyctiphanes australis*), which are pale and have bioluminescent organs along their bodies that twinkle like stars, and **lobster kril** (juvenile squat lobsters, *Munida gregaria*), which are a deep red colour with tightly curled tails and prominent claws, form dense bands along the tideline every now and then.

At higher latitudes – the Antarctic and some sub-Antarctic islands – shoals of **Antarctic krill** (*Euphausia superba*) can also occasionally become stranded on shore. These are the largest krill, reaching four to six centimetres in length. They normally hang out hundreds of metres below the surface during the day, rising up to shallower waters at night to feed on tiny algae (phytoplankton) – especially algae growing in sea ice.

Krill are a key food source for many iconic marine seabirds and mammals in southern waters, including baleen whales and penguins, but populations have declined drastically in recent decades, partly because of changes in sea-ice cover, and partly because humans harvest krill for health supplements.

The hyperiid amphipod *Themisto gaudichaudii* is abundant and widespread. Millions sometimes wash up on beaches in mass strandings.

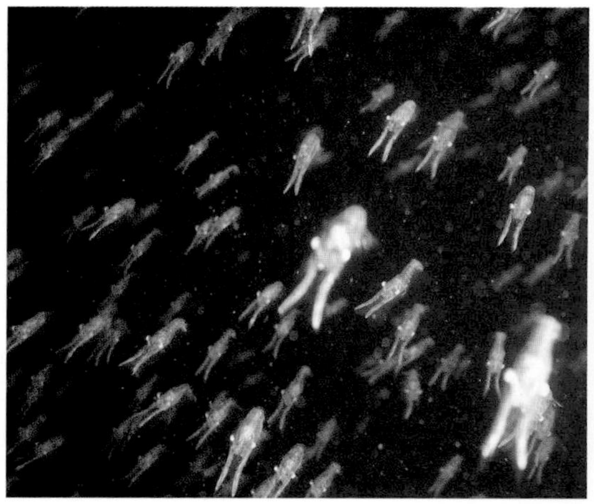

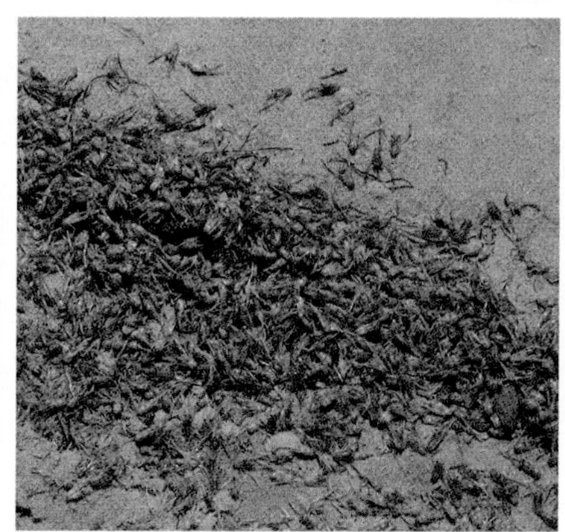

CLOCKWISE FROM TOP LEFT: Postlarval squat lobster *Munida gregaria* in New Zealand.

Shoaling postlarval squat lobsters (lobster krill) swim near the Falkland Islands.

Munida gregaria in a mass stranding in New Zealand. *Photo Jean McKinnon*

White krill, *Nyctiphanes australis*, on a New Zealand beach. *Photo Nick Beckwith*

There are many larger solitary decapod (10-limbed) crustaceans, such as **crabs** and **lobsters**, in shallow coastal waters. When they die, through predation, injury or disease, their bodies or pieces of their hard exoskeletons often wash up on shore.

Barnacles are crustaceans that grow a hard shell. Some species attach to rocks in the intertidal zone – the area between the high- and low-water marks – and you can sometimes find the detached shells of

CLOCKWISE FROM RIGHT:
A clump of acorn barnacle shells.

Part of a crab's outer 'shell' or carapace.

A dead crayfish.

The moulted carapace of a decorator crab.

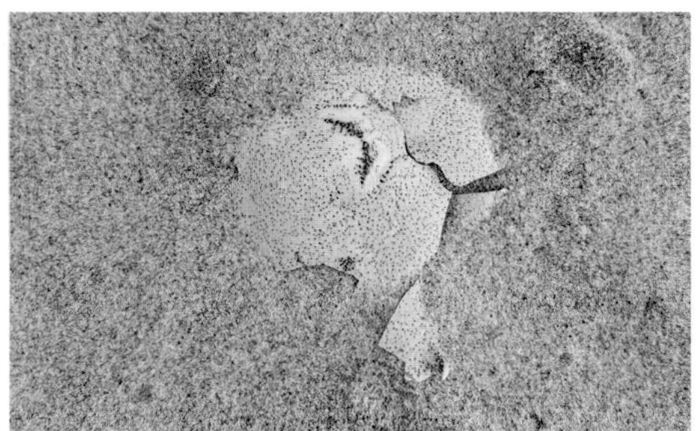

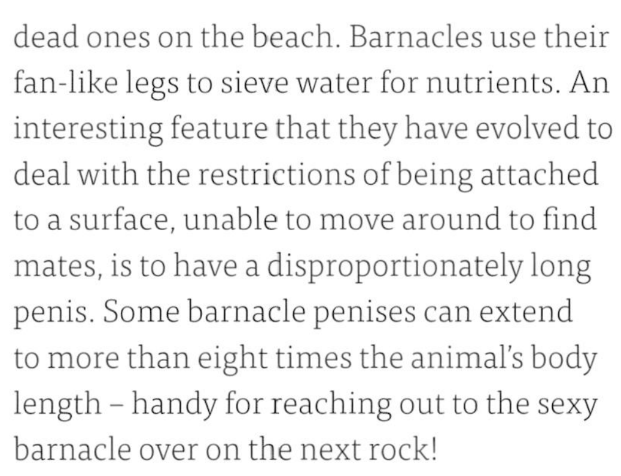

dead ones on the beach. Barnacles use their fan-like legs to sieve water for nutrients. An interesting feature that they have evolved to deal with the restrictions of being attached to a surface, unable to move around to find mates, is to have a disproportionately long penis. Some barnacle penises can extend to more than eight times the animal's body length – handy for reaching out to the sexy barnacle over on the next rock!

Bioluminescence can make oceans glow, especially where waves crash on the shore – a phenomenon caused by agitated dinoflagellates. Adventure Bay, Tasmania. *Photo Fran Davis*

BIOLUMINESCENT WAVES

Sometimes when coastal waters are warm, the waves will light up with a sudden electric-blue glow, flashing through the crashing waves. When you wade into the water it sparkles and zaps with light around your legs, and passing fish leave glowing trails. Glowing seas are caused by **bioluminescence** – light produced by living organisms. Most marine bioluminescence is blue, because the short wavelengths of blue light travel well through water.

Although many organisms in the ocean produce bioluminescence, when the ocean glows it is usually because of a dinoflagellate bloom – a population explosion. **Dinoflagellates** – tiny single-celled organisms that have whip-like strings (flagella) that they use to help them move through the water – are usually considered part of the phytoplankton, the drifting algae that form the base of most marine food chains. Many dinoflagellate species photosynthesise, getting their energy from sunlight just as seaweeds and plants do. Some species, however, also consume other organisms, for example through phagocytosis: wrapping plasma membrane around another small organism that is then incorporated into the dinoflagellate cell.

The dinoflagellate's colour can be partly the result of what it has taken in through this process – the algae that it holds in its fluid-filled cavities (vacuoles). Dinoflagellates also contribute to 'red tides', where phytoplankton bloom in vast numbers, turning the surface of the water red.

Many dinoflagellate species can bioluminesce, but the widespread *Noctiluca scintillans*, which can be found in all of the world's oceans, is one of the best-known and can be responsible for both red tides and bioluminescent waves. This species has tiny round organelles called scintillons in its cell, from which light is produced through a chemical reaction when the species is agitated (for example, when a wave or animal pushes it suddenly). This startle response is thought to be for predator evasion – the light could confuse a predator, or might attract secondary predators to catch those feeding directly on the dinoflagellate.

Marine bioluminescence is produced by a class of compounds called luciferins, and enzymes that bind to them. When the acidity of the cell is increased, the enzyme luciferase changes shape, allowing luciferin to bind and creating a short, sharp flash of light. Although there are many different forms of luciferin, the same one found in dinoflagellates also occurs in krill, perhaps because krill feed on phytoplankton.

SEA FOAM

The agitation of water by wind and waves can cause **sea foam** to form. Seawater contains many particles of organic material, such as fats and proteins from the algae and animals that live in the ocean. Algal blooms, waves crashing through kelp forests, or surf stirring up sediments all lead to higher than usual amounts of these organic materials in coastal waters, and can result in the formation of sea foam.

Sea foam is common around the world and is usually not dangerous, although in some cases it can include toxins, for example from human pollutants. When foam gets particularly thick it can cause visibility problems for surfers and swimmers.

Sea foam forms when surfactants from seaweeds and other sources are agitated by wave action.
Photo Keith Probert

Ambergris often resembles a stone, but this valuable material is produced in the gut of a sperm whale. *Photo Lloyd Esler*

Beach Treasure

AMBERGRIS

Ambergris is – or was – one of the most valuable things you could find on a beach. A lump of good-quality ambergris was once worth tens or even hundreds of thousands of dollars, and could still be quite valuable, although these days synthetic substitutes have greatly reduced the price of the natural material. The value comes from its qualities as an agent for intensifying and preserving fragrances, in particular for helping perfumes to retain their smell for a long time.

Ambergris comes from the intestines of the **sperm whale**. Exactly why some sperm whales (1–5 percent, and mostly male) produce ambergris is not well understood, but it's thought the thick excretion might help to protect the whales' guts from the sharp beaks of the squid they feed on. Normally whales regurgitate squid beaks, but sometimes these hard, indigestible parts of their prey pass through into the intestine and make their way to the rectum, where faecal material sticks to them. The lump that forms is stewed in a warm, bacteria-rich environment for a long time and, if it doesn't grow too large, is excreted when the whale defecates.

Most ambergris is recovered from whale carcasses, but it is possible to find lumps on the beach. Ambergris can look like a boulder or a pebble, or a piece of flotsam that's been drifting at sea for a long time. It is usually covered with a whitish crust, has a rounded or irregular shape and is rock-hard. As ambergris varies a great deal in shape, size and texture, it can be difficult for a non-expert to identify. The best test is to heat a piece of wire in a flame and poke it a few millimetres into the ambergris. A tacky, resinous black liquid that stretches out into sticky strings, like molasses, will form around the wire.

MINERALS

Sand is essentially small grains of minerals. The geological definition of sand is mineral particles with a diameter between about 0.06 and 2mm – smaller than that and it's silt; larger than that and it's gravel. A single cupful will contain millions of grains. If you take a bit of the beach home and look at it under a microscope – or carry a good magnifying glass with you to the beach – you will find an astonishing diversity of colours, shapes and patterns in the sand.

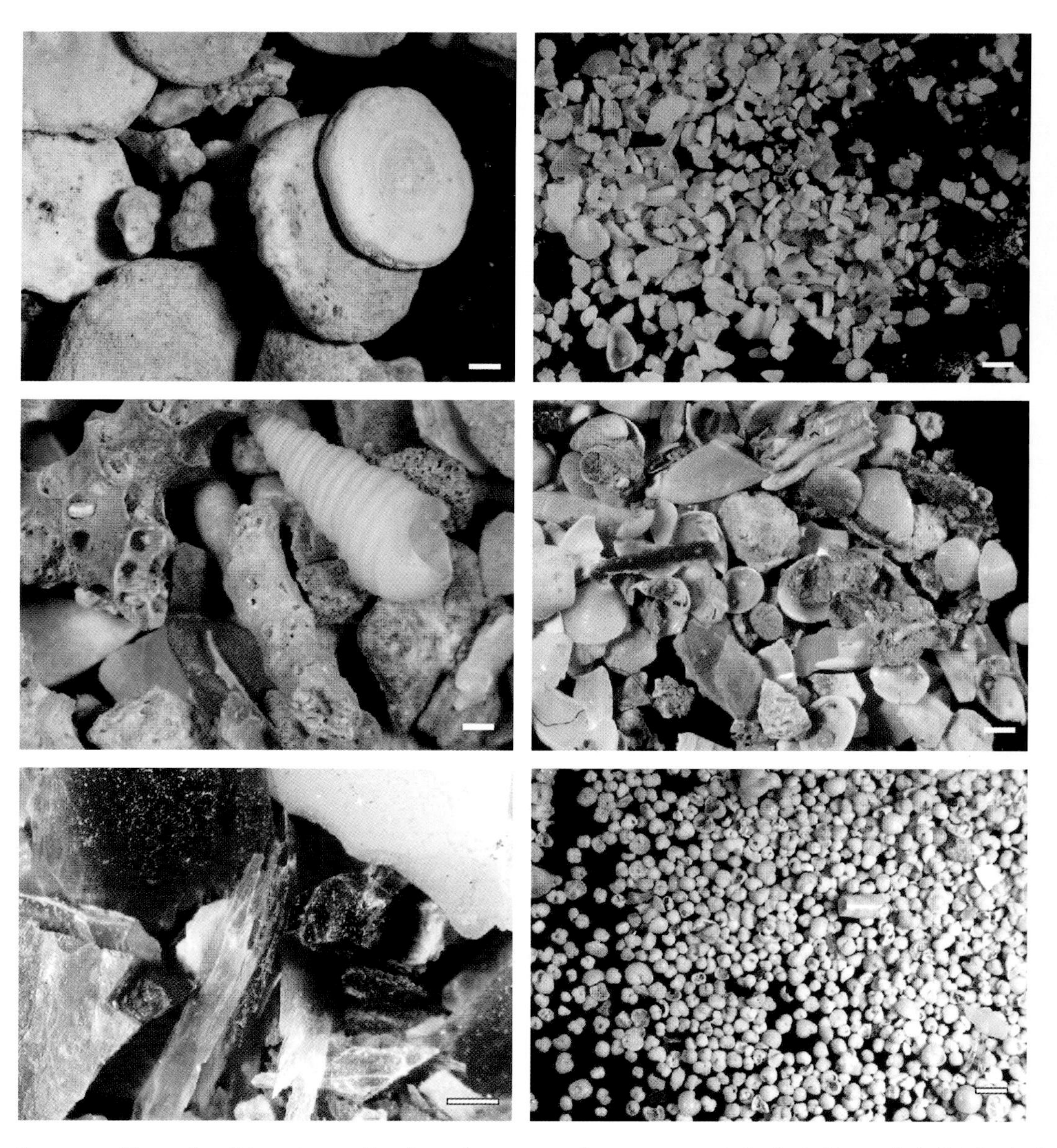

Some sand is made of tiny pieces of rock, such as quartz, but many sands include biogenic (of biological origin) materials, such as the frustules of diatoms, tests of forams, and crushed shells.

These magnified images (scale bars ~1mm) of sands (right) and larger-grained marine sediments (left) from New Zealand and Australia show some of the diversity of grain types found in biogenic sediments.

Every beach is a little different. Some are sandy, some are pebbly, some are mostly coral rubble, and some consist of many small shells or shell pieces. Most beach sand will include at least some diatom or foraminifera shells: tiny single-celled organisms that drift around in the ocean and form the basis of many food chains. Diatoms photosynthesise – they get energy from sunlight – and make their shell (their 'frustule') out of silica. These shells are like intricate glass ornaments. Foraminifera, on the other hand, don't photosynthesise – instead, they get their energy from organic matter in the water. Some make their shell (their 'test') out of calcium carbonate, while others bind together grains of sand to form a shell.

Ocean-tumbled glass fragments become rounded and resemble colourful river or beach pebbles.

A lot of sand is made of rounded fragments of rock, created as waves nibble away at cliff faces and boulders. Most of these sand grains are light-coloured quartz and other silicate minerals, but a few are metallic minerals that are more dense and create local concentrations of black sand. Black sand is mined for iron in some places, and many black sand deposits contain other valuable resources such as minerals with rare earth elements that are used in modern electronics. Small amounts of platinum (often as an iron alloy) and gold can be found in black sand, especially near where rivers wash the minerals out to the coast. Gold and platinum can be recovered by panning the black sand; the grains are usually so fine that they float on the water surface during panning.

Quartz-rich sand is high in silica and can be melted to make glass; glass-making techniques were discovered over 6000 years ago in the Middle East. Glass also forms naturally as obsidian when silica-rich lavas erupt from volcanoes, or as fulgurites when sand is struck by lightning. On many beaches you will find pieces of sand-tumbled human-made glass, slowly returning to their origin.

Quicksand can be a hidden danger, especially after heavy rain. When sand or silt mixes with water it can become saturated to the point that it becomes a non-Newtonian fluid – it appears solid but can change to liquid when only a small amount of force, such as a footstep, is applied. A human would not normally sink fully in quicksand, but panicked thrashing movements can increase sinking. If caught in quicksand, move your legs slowly to keep the sand liquid around you, and try to shift gradually into a more horizontal position.

Anyone who has ever lost their keys, coins or a ring on a beach knows how quickly the sand can cover all traces of small objects, and how hard they can be to find. Sometimes opportunistic treasure-seekers with metal detectors may be seen wandering along busy tourist beaches at the end of the day in search of treasure.

You might find **rocks** on a beach that don't seem to come from any local source. Scientists have found that floating kelp can carry chunks of rock, attached to holdfasts, for hundreds of kilometres at sea. In the past, too, some ships used rocks as ballast, and these could be cast on shore if the ship were wrecked. Some basalt cobbles found on beaches in the far north of New Zealand are thought to have come from the wreck of the *William Denny*, a steamer carrying cargo to Sydney from Auckland that ran aground in fog at North Cape in 1857.

Bouldery beaches and coastal rock platforms can be great places to find fossils. **Fossils** occur in sedimentary rock that forms as layers of sediment build up over time. The lower layers become compacted by the pressure of overlying layers (lithification). If an animal or plant dies in an environment such as a beach or a silty estuary and is rapidly covered by sediment layers, the hard parts of the body can petrify (become stone). In petrification, minerals such as silica or iron seep into spaces created by the organic remains, such as the pores in bones and shells, and are deposited there as solids.

Many cliff faces and rock platforms are formed from sedimentary rocks that have been raised up by one or several of a range of geological processes. As they weather away with the wind and waves beating at them every day, they contribute sand to the beaches – and can also reveal their hidden secrets. Keep an eye out as you walk around and you could spot shell or even bone fossils in broken or weathered boulders. Sedimentary rocks are softer than igneous rocks (the latter are created by heating and cooling minerals, for example in volcanic eruptions). The shades of different layers that you see in sedimentary rock represent old beach surfaces.

Large rocks can travel at sea attached to buoyant kelp. *Top photo Jonathan Waters*

ABOVE: Cunjevoi grow in the intertidal zone but can appear on beaches if they are detached by the activity of fishers or other disturbances.

From the Shallows

CUNJEVOI AND SEA TULIPS

Tunicates – many of which we call **sea squirts** because they squirt water – are tube-like animals that normally attach to rock or other surfaces. Each animal has an 'in' and an 'out' tube (the oral and atrial siphons, respectively); they suck water in through the oral siphon and sieve particles out of it through a mucus-lined basket (which really looks a lot like a basket!), before expelling the water through the atrial siphon.

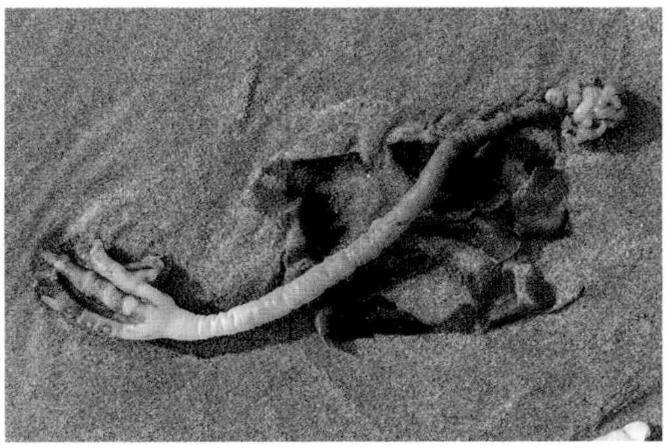

LEFT AND ABOVE: Sea tulips are stalked ascidians (sea squirts), which are filter-feeding animals closely related to vertebrates.
Top photo Kerryn Wood

Tunicates belong to a large phylum (biological group) with many subgroups, including the class Ascidiacea (ascidians), which have a tough cellulose tunic – the outer layer of their body. Most of those you commonly see on the beach are ascidians.

Pyura is a large ascidian group that includes the **cunjevoi** or 'red bait' species found on intertidal rock platforms throughout the Southern Hemisphere. Cunjevoi are not strictly all one species but are similar and closely related, so are grouped under the name *Pyura stolonifera*. These squat ascidians usually occur in dense aggregations that block the establishment of seaweeds, barnacles and molluscs, but at the same time form a three-dimensional habitat, like a forest, which can provide shelter for many other species. They tend to be covered in small green or red algae and grow so close together that some may be attached to others. They sometimes wash up on the beach in small groups. The flesh of ascidians is often used for fishing bait.

Some ascidians – including some *Pyura* – have long stalks and resemble flower buds; these are often called **sea tulips**. They are common in shallow waters and wash up on beaches after storms or when the sea floor has been disturbed.

Tunicates are more closely related to us than you might think. Many species produce free-swimming larvae that look a little like tadpoles, with a long tail and a notochord – a tough rod that, in vertebrates, becomes the backbone as the animal develops. Normally young ascidians lose their tail and fins when they settle down and develop into adults, but scientists think this free-swimming stage could have given rise to all other chordates – animals with a notochord – including vertebrates (everything with a backbone). That evolutionary leap might have come about if some tunicate larvae were able to reproduce before or without entering the attached adult part of the life cycle, remaining as swimming tadpole-like creatures.

SPONGES

We all know what a sponge feels like from the synthetic versions we buy for cleaning, but before polyurethane foam began to be produced in the 1940s and 50s, people used natural sea sponges. Ancient Greeks and Romans sometimes dipped sponges in fragrant oils and used them to clean and perfume their bodies.

Sponges (the non-synthetic sort!) are animals in the phylum Porifera (meaning 'having pores'). They are filter feeders, but unlike tunicates they don't have a gut, a nervous system or a heart. Instead, their bodies are essentially sieves that filter organic particles from water.

There are three groups of sponges: demosponges, which make up around 90 percent of sponge species and which have tiny hard parts (spicules) of either calcium carbonate or silicon dioxide, usually in a protein matrix, made of collagen and called spongin, that gives them their ... er ... sponginess; 'glass' (hexactinellid) sponges, which have spicules made only of silicon dioxide fused together in a lattice; and calcareous sponges, whose spicules are made of calcium carbonate.

The latter two groups do not have the spongy protein and are generally more brittle than demosponges. Glass sponges are mainly found in the deep sea or polar regions, so you are less likely to find these washed up on a beach.

Demosponges form many different shapes, even within species. Some are tube-like or cup-shaped while others are bulbous, branching and tree-like or thin and wavy, like fans. The shapes are partly driven by the environment – for example, the form might change depending on the size of the sponge and how fast water flow is in the area (some shapes are stronger than others).

Sponges come in a wide range of shapes, sizes and colours. *Above photo Mia Ching*

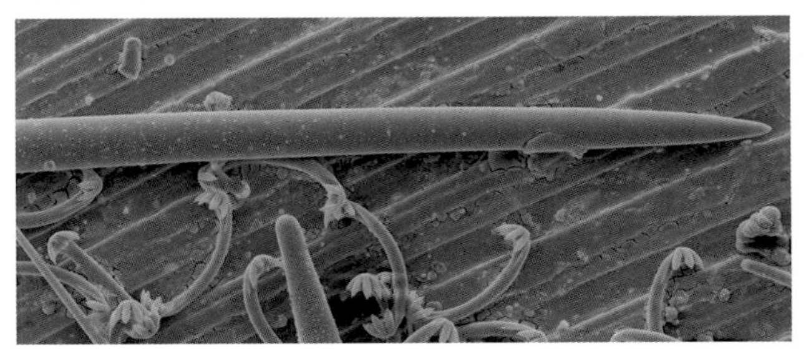

RIGHT: Spicules – the hard internal parts – of a sub-Antarctic demosponge, *Chondrocladia*. The smaller hook-like spicules (less than 1mm long) help the sponge capture small prey. *Scanning Electron Microscope image: Rachel Downey*

Individual sponge cells perform different roles, such as for immunity, reproduction, spicule secretion, collagen secretion and digestion. Astonishingly, these cells can change their roles; as a result, if a living sponge is damaged it can quickly recover by deploying cells into new roles. The cells of a sponge that has been torn apart, for example in a food processor, can even reaggregate into a functional sponge!

Sponges create three-dimensional habitats for many other organisms. Inside a sponge you might find crustaceans such as amphipods, isopods and crabs, as well as algae, worms and echinoderms, especially brittle stars.

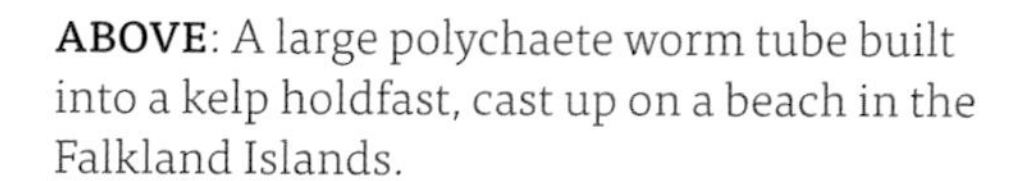

ABOVE: A large polychaete worm tube built into a kelp holdfast, cast up on a beach in the Falkland Islands.

TOP RIGHT AND MIDDLE: Amphinomid polychaetes (or 'sea mice') found on beaches in New Zealand. *Photos: Top Nick Beckwith; middle Chris Woods*

RIGHT: A cone-shaped pectinariid polychaete worm tube made of sand grains stuck together with mucus. *Photo Chris Woods*

WORMS

There are many different marine worm species, some of which live in or near beaches. Many marine worms are **polychaetes**, meaning 'many hairs', and indeed most have bristles along their sides coming out of fleshy protrusions ('parapodia' or legs), which can make them look a little like centipedes. They use these bristles in a range of ways, including for movement. Some have scales and others have long flowing tentacles that they use for feeding. Some have spiralled or fan-like structures on top that look like palms or Christmas trees, which they use to filter food from the water and to get oxygen; others have strong, hard jaw parts for catching and killing prey. Some build elaborate and often beautiful tubes out of mud, sand or shells bound together with mucus. You are more likely to find the vacated tubes of worms lying on the beach than the worms themselves, as the worms are tasty treats for fish and birds when exposed.

Beachworms are common on surf beaches and are generally predators; in the Southern Hemisphere (particularly in southern Australia) most are from the family Onuphidae. If you're familiar with the movie *Dune*, you've already seen marine beachworms from the perspective of their prey. Some species are skilled ambush hunters: buried in the sand in burrows, they emerge when something delicious walks, crawls, slides or swims past. Many people use these worms as bait for fishing – they entice them by dragging nets of smelly fish parts along the sand and slowly pulling the emerging worms from their burrows. Most polychaetes have larvae (young) that drift at sea, and genetic studies have indicated that populations along coastlines can be quite well connected, with larvae apparently able to disperse via ocean currents to different beaches.

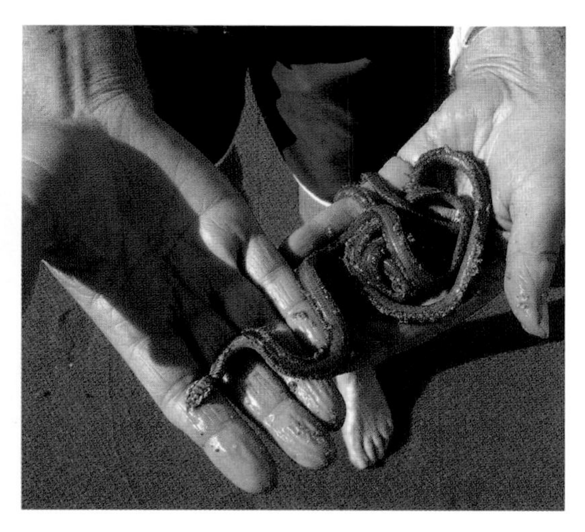

ABOVE: *Perinereis* worms are large predatory polychaetes that are often found under rocks in the intertidal zone and shallows. When alive or fresh they have bright, iridescent colours.

LEFT: An onuphid beachworm (a 'kingworm') from eastern Australia. *Photo Hannelore Paxton*

The other main group of worms that you are likely to come across on a beach or intertidal rock platform are the **serpulid polychaetes**. These worms make hard, white, calcium carbonate tubes attached to

Serpulid worm tubes on an intertidal stone in New Zealand.

The gelatinous egg mass of a moon snail. These horseshoe-shaped blobs are sometimes mistaken for sea jellies. Photo Kerryn Wood

rocks, shells, wood or other solid objects. The serpulids are filter feeders, using fan-like crowns or tentacles to create water currents that cause small food particles to become trapped in their mucus and to slide into their mouths. They will grow on buoyant objects drifting at sea, and are often cast up on beaches.

A hydrozoan medusa – *Aequorea* (a type of sea jelly) – washes in to shore. Photo Chris Woods

JELLY BLOBS

Clear, gelatinous masses that wash up on beaches can be one of a few things. Perhaps the most common are predatory **snail egg masses**. If you look very closely and see small dots inside, those are the eggs, held in a jelly matrix. Snail egg masses such as these are usually not circular. They are much larger than the snail that laid them, as the jelly absorbs water and expands. After a few days it breaks up and releases larval snails. The eggs on the outer parts of the jelly blob tend to develop most rapidly because they have better access to oxygen diffusing through the matrix; when they start to hatch this improves the oxygen transfer to the inner eggs. The larvae are planktonic (they drift in the water column) until they develop enough to become benthic (based on the sea floor).

Gelatinous salps sometimes wash up on beaches.

In contrast, if you find a circular jelly blob on the beach and, on closer inspection, notice concentric rings and radial patterns, this is probably the 'bell' of a **sea jelly** that might have lost its more fragile tentacles. Sea jellies are also known as jellyfish. (The marine groups that used to have 'fish' as part of their name, but which are not really fish, have been given new and more correct common names: starfish and jellyfish have become 'sea stars' and 'sea jellies'.)

Other gelatinous blobs found on beaches could be **salps** – transparent tunicates that are bell- or barrel-shaped, and which live out in the open ocean. They feed on plankton (tiny plants and animals, including larvae, that drift through the sea driven by currents), and move by jet propulsion by contracting to squeeze water out of their bell and squirting forward. Salps are either solitary or joined together in chains, depending on the stage of their life cycle.

Comb jellies (ctenophores, also known as 'sea gooseberries') are not sea jellies, although they look similar. They also resemble salps in that they have barrel-like bodies, but they have rows of cilia (thin hair-like structures) running along their bodies, which they use to move through the water. When viewed underwater, the pulsing motion of these wiggling cilia scatters light into a rainbow of colour. When washed up on a beach, comb jellies are less spectacular and might be hard to distinguish from other gelatinous blobs, but the rows of fused cilia should be visible as ridges or tucks running along the animal from top to bottom. Comb jellies catch prey with sticky rather than stinging tentacles.

Anemones growing together attached to a rock in a shallow subtidal pool.

A dead wandering anemone on a stony beach in North Canterbury, New Zealand.

ANEMONES

Anemones are cnidarians, related to sea jellies and corals. They use stinging tentacles to catch prey such as small fish.

Most attach to rocks, in rock pools or shallow water, and rarely wash ashore, but the **wandering anemone**, *Phlyctenactis tuberculosa*, which occurs around southeastern Australia and New Zealand, moves about by creeping or drifting and will sometimes turn up on a beach. This odd-looking creature resembles a ball of baked beans during the day, when it is curled up; at night it uncurls to show its tentacles. The wandering anemone can often be seen clinging to seaweeds in shallow water. Dead and dried on the beach they look a little like corals but are lighter and have a spongy feel.

A live wandering anemone clings to the seaweed *Hormosira banksii* (Neptune's necklace) near Kaikōura, New Zealand. Photo John Sullivan

SEAWEEDS

Beach-cast seaweeds play critical roles in near-shore ecosystems, moving nutrients from the ocean onto the land. Many invertebrates such as insects and crustaceans, as well as some vertebrates, feed on beach-cast seaweeds, and nutrients from decomposing seaweeds nourish animals that live below the surface of the sand.

Seaweeds get their energy from sunlight through photosynthesis. They are all algae, and to differentiate them from microscopic algae such as diatoms, they're referred to as macroalgae. Seaweeds are different to land plants, which use roots to get nutrients and water from soil. Although seaweeds have a base that can look a little like roots, this 'holdfast' is only for attachment. Nutrients are absorbed directly from the water through the seaweed's fronds.

Seaweeds are broadly split into three groups: green, red and brown (Chlorophyta, Rhodophyta and Phaeophyceae). You can usually – although not always! – recognise which group a seaweed belongs to from its colour.

Seaweeds can help to make sand. Some green and red seaweeds are calcareous, which means they create calcium carbonate 'skeletons', and when these skeletons break down they contribute to sedimentation, turning into sand. **Coralline algae** are calcareous red seaweeds that often look like pink lichen, either forming flattish crusts (nongeniculate, crustose algae) or branching like trees (geniculate corallines). Coralline algae appear to play important structural roles on rocky and coral reefs. For example, in 2016 a 7.8-magnitude earthquake lifted parts of New Zealand's east coast by up to six metres. Over the following two to three years, the gradual disintegration of the coralline algae crusts on uplifted rocks contributed to widespread erosion of the newly exposed seabed.

Crustose coralline algae (flat paint-like pinks on lower rock) and geniculate coralline algae (frilly pinks on upper rock), New Zealand.

5 days

5 months

3 years

Following an earthquake that uplifted subtidal rocks by some six metres, the decay of the hard, crusting coralline algae coating led to a loss of rock structural integrity and widespread erosion. The coralline algae looks like pink paint in the left photo, and – when dead – like white paint in the central and right photos. Waipapa Bay, New Zealand.

Red seaweeds from the genus *Gigartina*, which can form narrow or wide sheets covered in spiky, pimply bumps, often wash up on southern beaches. *Gigartina* extract is available as an anti-viral health supplement, and *Gigartina* is one of a few seaweeds that produce carrageenans instead of agars. Carrageenans and agars are polysaccharides (carbohydrates) found in seaweeds that are important gelling agents and are used in laboratory applications and across many products, such as in toothpaste, ice-cream, milk, yoghurt, salad dressing and cake icing, to thicken liquids or keep particles in suspension. Several countries in South America have large seaweed export and/or agar production industries. The branching red seaweed genus *Gelidium* is a particularly important source of agar in South America.

Common **green seaweeds** that you might find washed up on Southern Hemisphere beaches include the genera *Ulva*, with wide lettuce-like fronds or long stringy fronds (the stringy species, *Ulva intestinalis*, used to be in the genus *Enteromorpha*); *Codium*, which can be branching or cushion-like; and *Chaetomorpha*, which looks like a string of emeralds.

In Chile seaweeds are collected for agar, often in cottage-industry style. People use poles to pull seaweeds from the shallows; these are then air-dried and sold to local agar factories or exporters.

A detached *Gigartina circumcincta* in southern New Zealand. The green seaweed in this image is *Ulva lactuca* (sea lettuce), and the pale pinky-white lace-like seaweeds are geniculate corallines.

Codium fragile in Tasmania. This green seaweed has become invasive in some areas.

A huge pile of giant bladder kelp, *Macrocystis pyrifera*, washed up on a beach in Otago, New Zealand.

Most beaches will, at some time, be dotted with washed-up large **brown seaweeds**, which most people think of as kelp. Some scientists prefer to save the word 'kelp' only for one particular group of large brown seaweeds (Laminariales), but the term originally also included other groups, such as the Fucales, and is a common word with wide usage.

Kelp species come in a great variety of shapes. These brown algae are ecologically important, forming both habitat and food for many invertebrate and vertebrate species including crustaceans (such as crabs, amphipods, isopods and crayfish), echinoderms (such as urchins and sea stars) and fish. Many kelp species are fast-growing, so can rapidly sequester carbon: *Macrocystis pyrifera*, or **giant bladder kelp**, is the fastest-growing organism in the world and can grow up to 60cm per day. This giant kelp is found throughout the cool-temperate parts of the world. Although its populations are, in the Southern Hemisphere, shifting south and declining in many places as waters warm, it is often found washed up on beaches.

CLOCKWISE FROM RIGHT:
Giant bladder kelp grows in tall subtidal forests. This species floats well so is often cast up on beaches, sometimes far from where it grew.

The holdfasts of many kelp species form ideal 'cities' for small animals to live in.

Some kelp holdfasts can reach enormous sizes, such as this *Macrocystis* holdfast which is over a metre wide and washed up on a beach in the Falkland Islands.

The bladders or pneumatocysts of giant bladder kelp are gas-filled structures that act as buoys, holding the kelp upright in the water and keeping the fronds as close as possible to the sunlight at the surface.

The soft fronds of this giant bladder kelp have been torn off by wave action, leaving only stipes (stalks) and pneumatocysts.

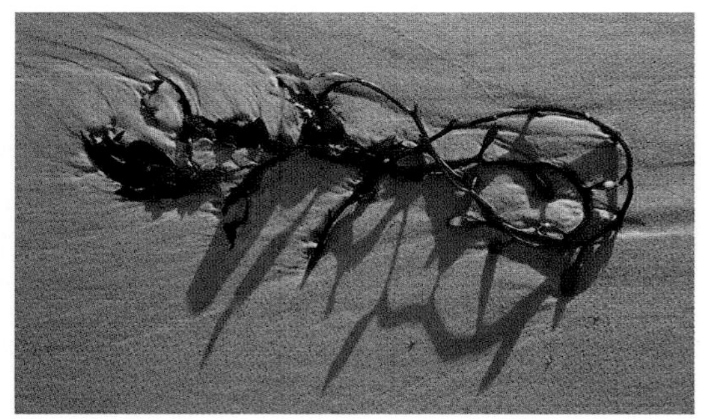

Southern bull kelp (*Durvillaea*) grows throughout much of the cool-temperate and sub-Antarctic regions of the Southern Hemisphere. Most species are found in the New Zealand region, but Chile and Australia also have species that are found nowhere else. One species, *Durvillaea antarctica*, grows throughout the sub-Antarctic region and in New Zealand and southern Chile. This guide is reproduced from Fraser et al. 2020 (see Further Reading).

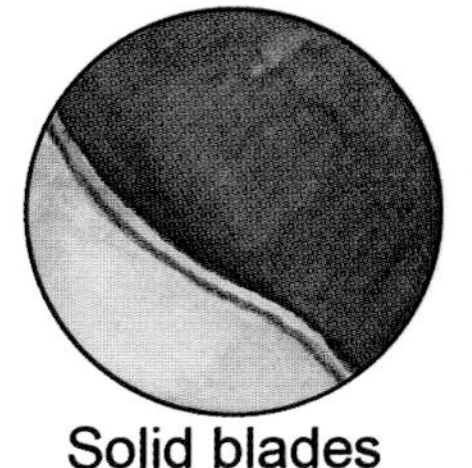

Solid blades
with rough, pale margins

Dense holdfast
not hollowed out by animals

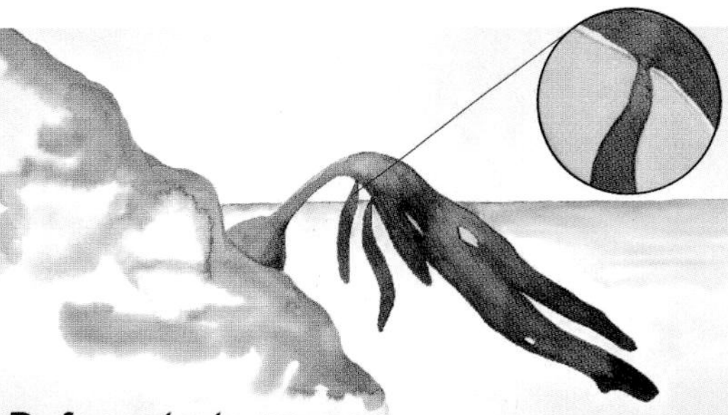

Lateral blades constricted at base, irregular placement.

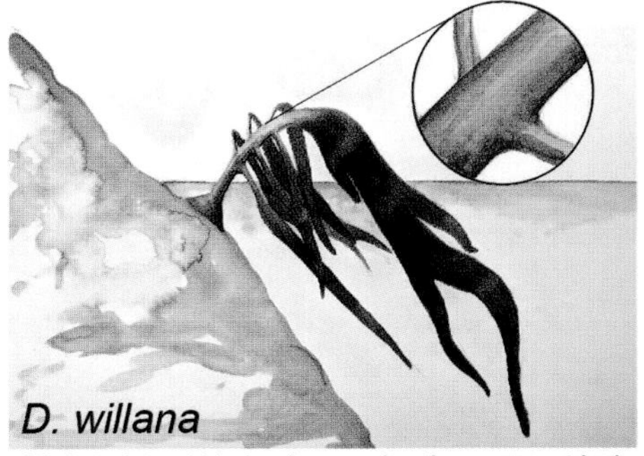

Stipitate lateral blades from main stipe, symmetrical.

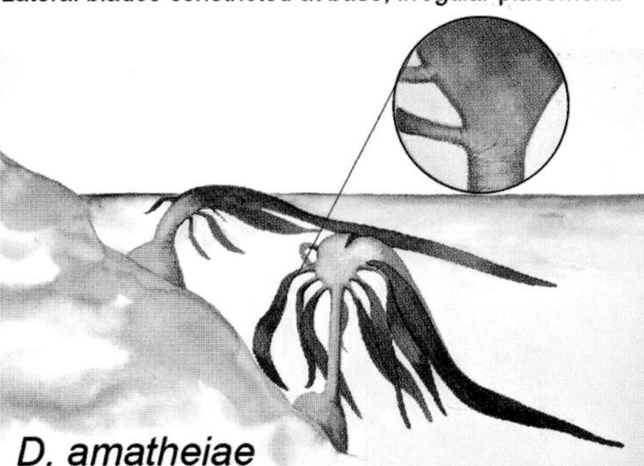

D. amatheiae
Stipitate lateral blades; irregular placement.
Relatively short (total length).

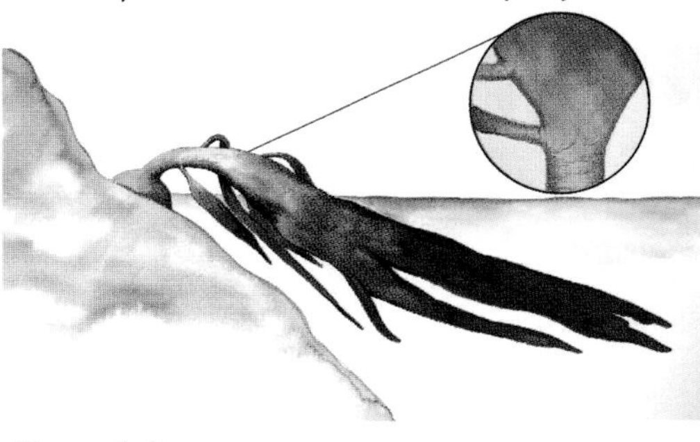

D. potatorum
Stipitate lateral blades; irregular placement.
Relatively long (total length); stipe base stout.

In southern Australia, New Zealand and southern South America, the large **southern bull kelp** *Durvillaea* often dominates intertidal and shallow subtidal ecosystems, growing up to 12m long. In Australia there are two species, neither of which can float, and in New Zealand there are two species that float and three that don't. In Chile *D. incurvata* occurs north of the fiords, and *D. antarctica* is found in Patagonia. Both float well. All are often found washed up on beaches.

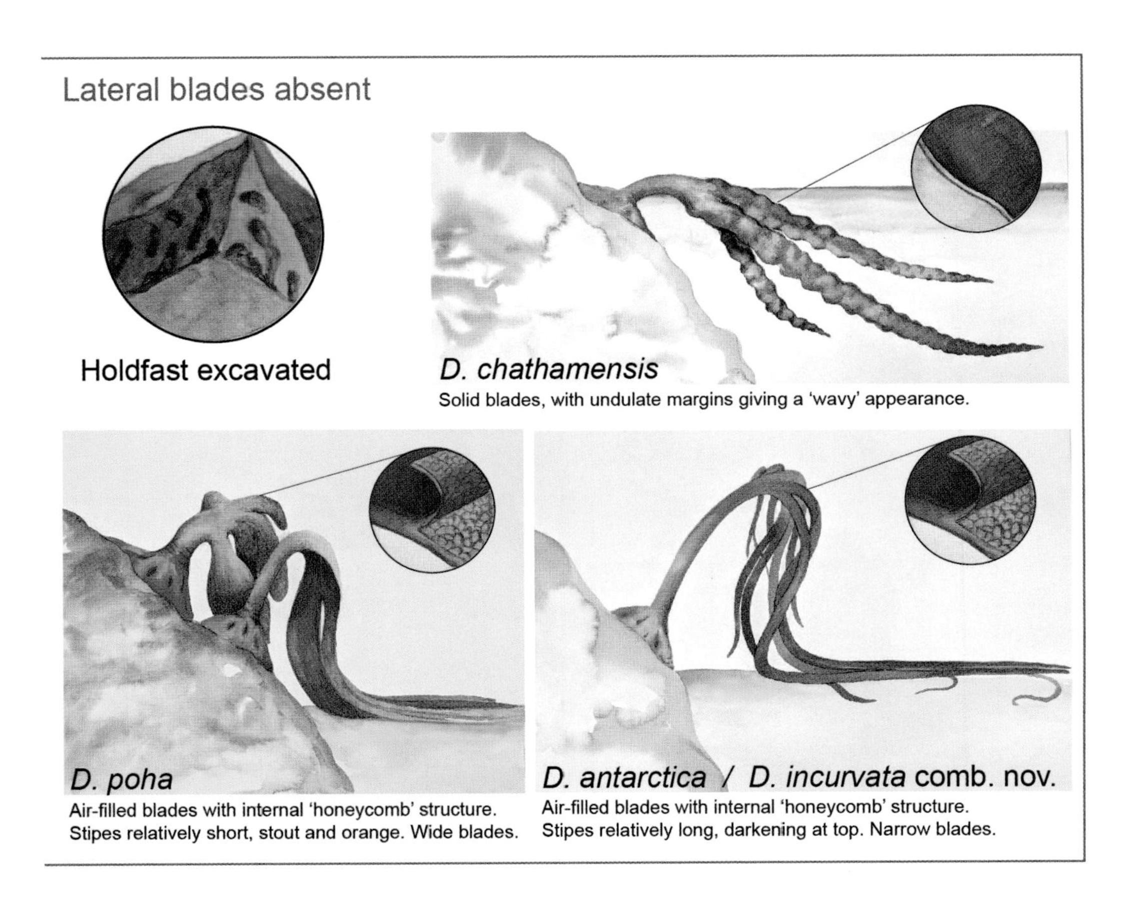

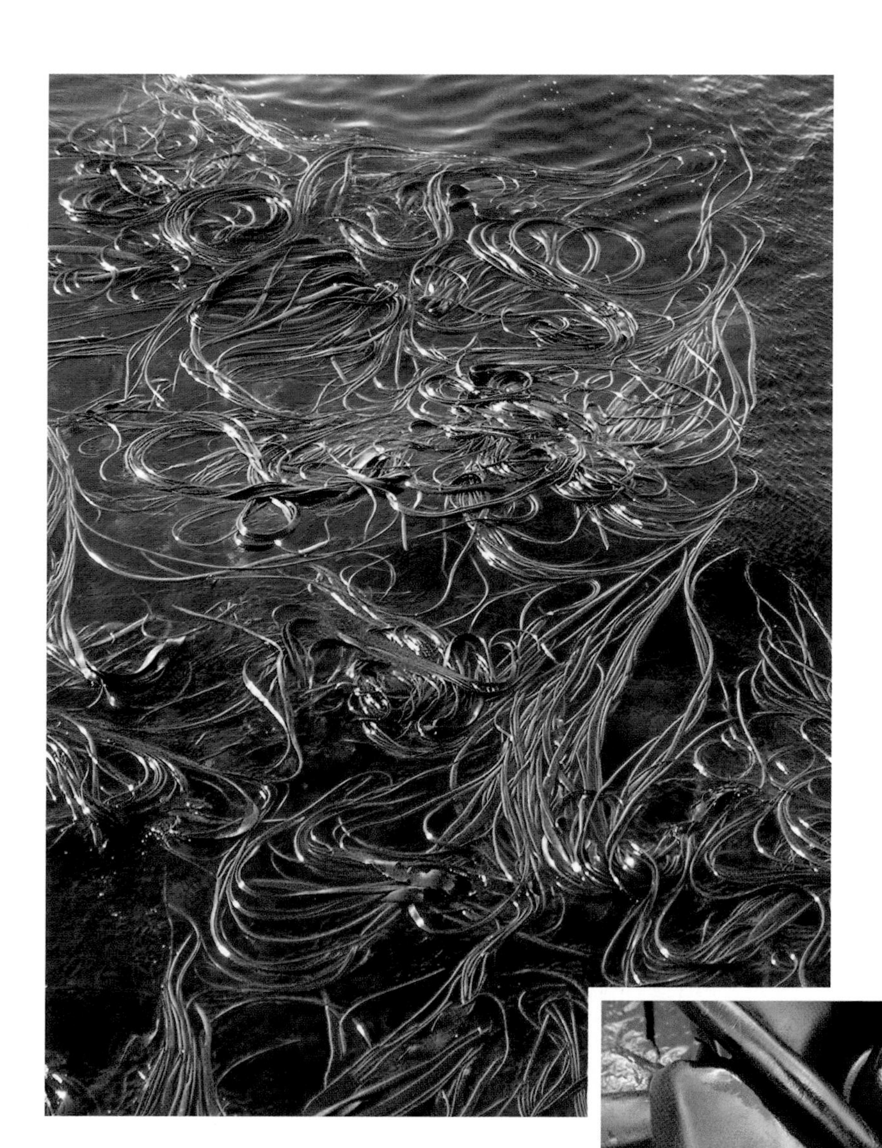

Southern bull kelp, *Durvillaea antarctica*, grows throughout the high latitudes of the Southern Hemisphere. The movement of its fronds resembles flowing hair, or spaghetti.

Ecklonia maxima washed up on a beach in South Africa, near Cape Town.

An unusual curly *Ecklonia maxima* stipe, South Africa.

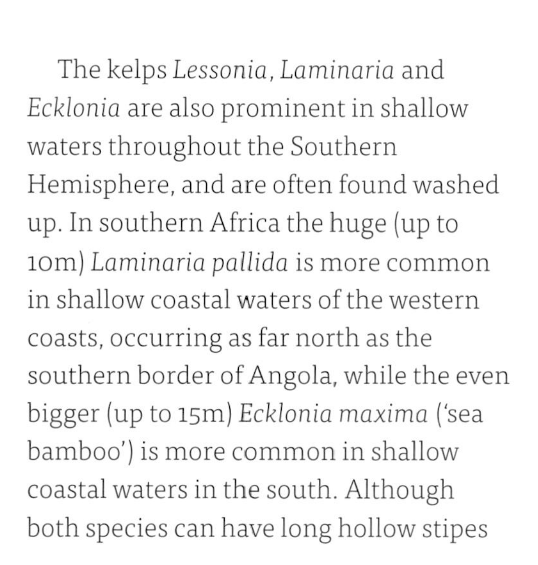

The kelps *Lessonia*, *Laminaria* and *Ecklonia* are also prominent in shallow waters throughout the Southern Hemisphere, and are often found washed up. In southern Africa the huge (up to 10m) *Laminaria pallida* is more common in shallow coastal waters of the western coasts, occurring as far north as the southern border of Angola, while the even bigger (up to 15m) *Ecklonia maxima* ('sea bamboo') is more common in shallow coastal waters in the south. Although both species can have long hollow stipes

Lessonia variegata in Chile.

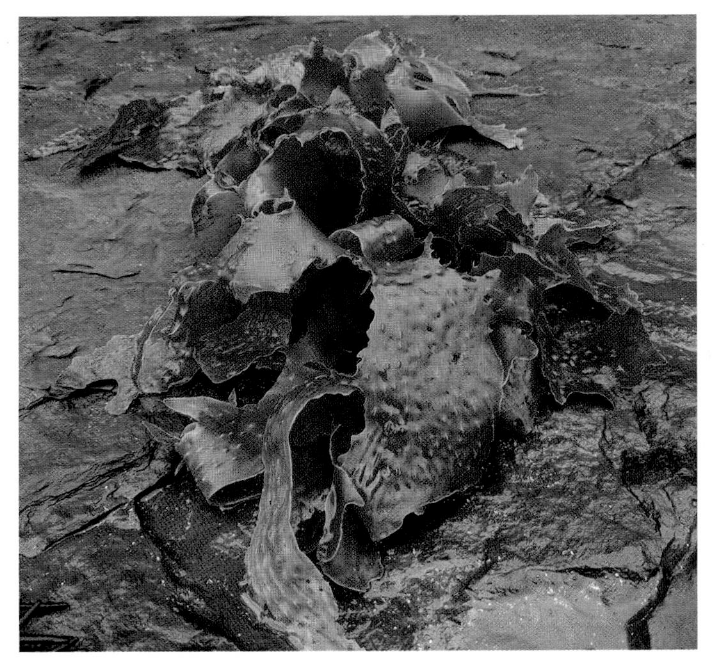

LEFT: *Ecklonia radiata* washed up on an intertidal rock platform in Tasmania.

RIGHT: *Marginariella in New Zealand.*

Hormosira banskii, 'Neptune's necklace', is a brown seaweed common in shallow waters around the Southern Hemisphere. This species floats well.

(the trunk-like part), *L. pallida* can be distinguished from *E. maxima* as its stipe does not bulge much at the top and it has a single fan-like blade divided into strips, whereas the blades of *E. maxima* seem to branch sideways off a central strip. The smaller *Ecklonia radiata* is also common and often washes up in Australia, New Zealand and Africa. This species has noticeable spiky growths all across its fronds and jagged frond edges.

Large kelps do not grow well in shallow coastal areas in the tropics, preferring cooler and more nutrient-rich waters.

Southern bull kelp (*Durvillaea*) is dried and sold for consumption at roadside stalls and supermarkets in Chile.

Almost all seaweeds are safe to eat as long as they are taken from clean waters (not, for example, near sewerage outlets or ports). Seaweeds are high in iodine, iron and vitamins. You don't need to consume huge amounts to get the benefits – and indeed, you can have too much of a good thing: some of the elements in seaweeds should not be consumed in excess ... and seaweeds do have a mild laxative effect! Soft thin seaweeds such as *Ulva* or *Pyropia* (previously *Porphyra*, or karengo in Māori) can be eaten raw and untreated – *Pyropia* is also dried in paper sheets (nori) for sushi and other Japanese dishes. Some others are more chewy and are best cooked or dried. Southern bull kelp is eaten in Chile but rarely straight out of the sea – it is stewed in seafood dishes, and dried pieces are added, like bacon or croutons, to salads and other meals or used as teething rusks for babies. Coralline seaweeds are not worth trying, however, unless you like eating sand!

Seaweeds have other applications as well. In New Zealand, Māori slice open the wide blades of *Durvillaea poha* to form air-tight bags (pōhā) to preserve muttonbirds and other foods in fat. Once sealed the bags kept the food safe for a long time. Cows on King Island eat *Durvillaea* washed up on beaches, for its salt and iodine, and beach-cast bull kelp is also commercially harvested there for alginates.

HOW TO COOK COCKLES IN KELP

In New Zealand or Chile, wide buoyant *Durvillaea* blades can be used to cook cockles or mussels. Simply slice through the soft, honeycombed internal parts to make a bag, stuff it with cockles and stitch the ends with twine. Place on a baking tray or over hot coals and cook at around 200°C for 15–20 minutes. The cockles will steam and adopt a seaweedy tang.

Photo Pamela Olmedo Rojas

KELP RAGOUT
CHARQUICÁN DE COCHAYUYO

Despite being tough and rubbery when fresh, bull kelp can be prepared in a variety of ways and is often eaten in Chile. A classic is kelp ragout.

Collect healthy fronds of buoyant *Durvillaea* (other large brown seaweeds could be substituted but ideally should have not-too-thick fronds). Boil fronds for around an hour in plenty of water. Allow to cool. Wash fronds under the tap to remove the remains of the toughest outer layers. Cut into small cubes.

Stir-fry ½ capsicum (diced), ½ cup onion (diced) and 3 cloves of crushed garlic in 2 tbsp olive oil, with a little cumin, oregano and chili powder (if available use merkén – Chilean smoked chili).

In a large pot, combine 700ml water with 3 cups diced pre-boiled kelp, 2 cups cubed potatoes, 2 cups cubed pumpkin, ½ cup peas, ½ cup corn kernels and ½ cup diced carrots. Bring to the boil and cook until vegetables are soft.

Add a bunch of spinach leaves and allow to wilt before serving.

Optional extras: serve with a fried egg, or add smoked meat while cooking.

Some seaweeds are ecologically problematic. The 'Asian' or 'Japanese' kelp *Undaria pinnatifida*, marketed as wakame when processed for food, is a highly invasive species that can outcompete native seaweeds in temperate waters of the Southern Hemisphere and has become established in Australia, Argentina and New Zealand. This large brown seaweed species can most easily be recognised by the delicate, almost slimy feel of its fronds, and by the wavy, ruff-like fronds along the base of its stipe.

Undaria pinnatifida has a ruff-like base. Blanched *Undaria* is used for wakame seaweed salad, often seasoned with sesame oil.

WAKAME SEAWEED SALAD

Soak kelp in fresh water for around 10 minutes; drain.

Chop kelp into strips, including thin parts of the central stipe.

Blanch with boiling water for a few minutes; drain.

Season with sauteed garlic, ¼ cup sake/white wine, 1 tsp sugar and 1 tsp sesame oil.

Sprinkle with toasted sesame seeds if desired.

Serve cold.

SEAGRASSES

Sometimes you can find, washed up on beaches, clumps or balls of detached grasses – not necessarily from the land.

Unlike seaweeds, which are algae, **seagrasses** are true plants: they have roots to take nutrients from the sediments. They are flowering plants (their flowers bloom underwater!) and, like seaweeds, are important in marine ecosystems, providing both habitat and food for a wide range of animals. Large creatures that feed on seagrass include turtles, dugongs and fish, and many fish species use seagrass meadows as nurseries – relatively safe places for their young to shelter. Seagrasses are fragile ecosystems; they are easily damaged by human activities and are in rapid decline around the world.

TOP RIGHT: *Zostera* seagrass cast up on a beach in Otago, New Zealand.

RIGHT: Detached *Posidonia* seagrass when rolled around by the ocean can tangle into balls. This also happens to some stringy seaweeds. *Photo Maria Capa*

ABOVE LEFT: Squid beaks are hard and sharp and often rejected by predators, so can wash up on beaches.

ABOVE: Cuttlebone, the internal hard part of a cuttlefish. *Photo Chris Wood*

SHELLS

There's something incredibly calming in searching for beautiful shells on a beach, a pastime that appeals to all ages. Shells are made by molluscs, but not all molluscs make hard external shells: for example, most squid and cuttlefish (which, along with octopuses, are the 'cephalopods', or head-foot molluscs) and sea slugs (nudibranchs) have relatively small, soft, internal shells.

Squid have a fine chitin 'pen' inside, as well as a hard beak. **Cuttlefish** have a larger, soft, layered internal shell called a cuttlebone – a lightweight oval plate of aragonite (a form of calcium carbonate) with internal chambers that the cuttlefish can use to adjust its buoyancy. Cuttlebones often wash up on beaches and are rich in calcium – you can buy them in pet shops as a dietary supplement for birds and reptiles. They are sometimes used by jewellers to make moulds, as they can cope with the high temperatures of molten metals.

Some cephalopods have structures that resemble curly snail shells. *Nautilus* and *Spirula*, for example, have spiralled, multi-chambered shells. Although both live at depths of several hundred metres, their shells have chambers that can fill with gas, making them buoyant and likely to wash up on shore when the animal dies. The shell of *Spirula* is internal, inside a fleshy mantle, and is used as a buoyancy device by the animal adjusting the amount of gas in its chambers. The shell of *Nautilus* is external, and the animal can retreat right inside its shell when threatened. *Nautilus* are like living fossils: some nautiloid fossils have been dated to hundreds of millions of years ago, well before dinosaurs roamed the Earth.

CLOCKWISE FROM ABOVE:
A *Nautilus* shell, broken along one edge.

A *Nautilus*.

When cut in half the *Nautilus* shell reveals its iridescent chambers.

Like *Nautilus*, *Spirula* is a cephalopod, but its shell is internal.

Two main groups of molluscs produce the hard, external shells that you most commonly find on beaches: bivalves and gastropods.

Bivalves have shells with two similar parts (valves) that swing open, joined with a ligament hinge. Some bivalves, such as **mussels** and **rock oysters**, are attached to hard surfaces and cannot move around. Others, such as **clams/cockles** and **scallops**, live in soft sediments, and some use their muscular 'foot' to move, or shoot short distances through the water by jet propulsion. Most bivalves are filter feeders, sieving organic particles from the water. Bivalves are often collected by humans for food, but their shells make an important contribution to sedimentation; over-harvesting can, if the shells are not returned to where they came from, reduce the formation of sand in some areas.

Sometimes you will see bivalve shells that have a single small hole drilled into them, as though someone has already prepared them for stringing on a necklace or wind chime. These holes may have been made by a boring **octopus** – which does not mean the octopus is a poor conversationalist! Rather, if the octopus can't prise the two valves of the shell apart, it will make a borehole, rasping with its salivary papilla (a fleshy mouthpart, the tip of which is lined with small teeth) and

Oysters attach one of their valves to rock.

Cockle and scallop bivalve shells.

The 'horse mussel', *Atrina zelandica* (above left), is a large bivalve with shells that can reach 30–40cm in length. Cockles (middle) and mussels are bivalves.
ABOVE RIGHT: Operculum (shell door) of the snail *Turbo petholatus*. Photo Toine Broekhoff

releasing a secretion that both helps to dissolve the calcium carbonate of the shell and contains a toxin that kills the animal inside. Boring through a shell might take an hour or two. Some snails, such as the New Zealand moon snail *Tanea zelandica*, also bore through bivalve shells.

Gastropods with external shells are commonly known as **snails**. Snails are univalves, as they have just one main shell part. Some species also have an operculum, like a plate which they use as a hard door to close the shell opening. Opercula are flat on one side with a spiral pattern; those of turban snails are colourful and sometimes called 'cats' eyes'. Many marine snails are good at surviving exposure to air and warmth and are common in the intertidal zone, which can get hot when exposed at low tide.

Gastropods come in a range of shapes from classic spiral snail forms (bottom) to flatter, open-based abalone (top) and limpets.

LEFT TO RIGHT: Gastropods: limpet shell; spider conch shell; shell of a *Semicassis* predatory snail.

Some marine snails are squat and rounded like garden snails, and others are tall, thin and conical, like ice-cream cones. They may have spiralled shells or simple, tent-like shells, such as the **limpet**, whose shell resembles a low-rise teepee or a hat with straight ridges radiating from a central point. Because limpets cannot retreat into their shell to protect themselves when pulled off a rock (as some other snails can), they're thought to have evolved in areas where predation levels were relatively low.

Abalones (family Haliotidae) form another group of gastropods that cannot fully retreat into their shell. The abalone's shell is a shallow spiralled dome with a series of respiratory pores along one edge and resembles the shape of a human ear. Like the limpet shell it is open on the underside. The inside of the shell is iridescent and patterned, the colours varying among species. The Australian **greenlip abalone**, *Haliotis laevigata*, has a pale creamy interior and its flesh is highly prized, especially in parts of Asia, where it may sell for hundreds to thousands of dollars per kilogram. As a result, poaching is a major problem in some areas. The abalone fishery in Australia is tightly controlled, and there have been several instances of criminal prosecution for abuse of the fishery rules and limits. Likewise, in South Africa, one abalone species (*Haliotis midae*) is considered a particular delicacy in Asia and poaching has become a considerable problem, compounded by corruption.

In New Zealand the best known species is *Haliotis iris*, or **pāua**. Pāua has blackish flesh on the outside and is not as highly sought after in Asian food markets, which has made control of the fishery (both recreational and commercial) in New Zealand less problematic than those

The shells of pāua abalone and Cook's turban snail (*Cookia sulcata*) are often found on beaches in New Zealand.

Some mollusc shells have an iridescent 'mother-of-pearl' interior made of nacre. Flat aragonite platelets form in thin (around 0.5 microns, or 0.0005mm) overlapping layers held together with elastic polymers to create an extraordinarily strong structure. Because the platelets overlap, it's difficult for cracks to spread through the shell wall. As the platelets are a similar thickness to visible light wavelengths (roughly 0.4–0.7 microns, from violets to reds), the various colours that make up white light are reflected from the nacre. As a viewer moves the object or changes their viewpoint, different hues reflect from the surface. Abalone and nautilus shells are particularly valued for their nacre. Photo Nick Beckwith

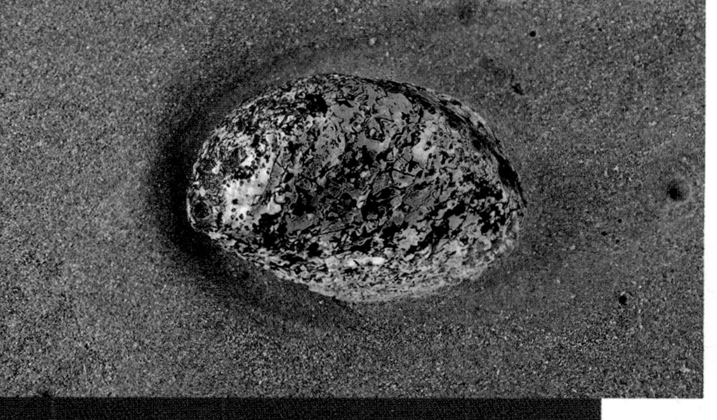

in Australia or South Africa. No licence is needed for recreational gathering of blackfoot pāua, although there are size and catch limits. The pāua's stunning iridescent-blue shells are prized for jewellery and other ornaments or are polished and sold whole. Creamy-coloured abalone shells found in New Zealand belong to *Haliotis australis*, **queen pāua**, and are smaller than the blue-shelled blackfoot pāua.

There is a huge diversity of gastropod species – over 62,000 have been named – but even within a single species there can be a large range of shell colours and patterns. Some of this variety could be driven by selection, since a snail's chance of survival is better if it can blend in with its background. Some colour selection might be driven by temperature, with dark tones that absorb heat well perhaps less favoured in hot places. The enormous range of colours and patterns can make browsing shells on beaches truly engrossing.

Another group of conical shells, appropriately known as 'cone shells' (genus *Conus*), can be found on beaches in tropical and subtropical areas. Whereas many conical snail shells have a small rounded opening at the wider end, the **cone snail** has a slit-like opening along its length, like a waffle ice-cream cone that is coming unrolled. Cone shells come in many patterns and colours and are prized in shell collections. However, it's important not to pick up a cone snail that is in the water as it might be alive. Living cone snails can be deadly: they are carnivorous and hunt for other animals, including vertebrates such as small fish. They have a venomous harpoon –

The tall conical *Maoricolpus roseus* in New Zealand and Australia has a great variety of colours and patterns.

ABOVE: Cone snails are venomous and their sting can be fatal, so do not attempt to pick up a live one. Their shells sometimes wash up on the beach. *Photo Imogen Fraser*

RIGHT: *Diloma durvillaea* snails cluster together to gorge on beach-cast southern bull kelp. *Photo Chris Woods*

BOTTOM: Hermit crabs use discarded snail shells as their homes, which they carry with them. *Pagurus albidianthus*, Otago, New Zealand. *Photo Jo Virens*

a barbed tooth that can be extended rapidly from the snail's proboscis, carrying toxin into the prey. Not all cone snails have toxins that are deadly to humans, but all can sting and are best left alone.

Hermit crabs sometimes use empty snail shells as elegant homes and protective cases. Hermit crabs have evolved for this purpose – their abdomen is coiled like the inside of a snail shell. When they grow too large for one shell, they choose another and move house.

Why can you hear the ocean when you hold some shells to your ear?
Snail shells – especially very large ones – are great sound amplifiers. When you hold one to your ear any background noise resonates, bouncing around the smooth hard walls of the chambers of the shell, and can sound like waves crashing on a beach. The bigger the shell, the deeper and more convincing is the ocean-like sound. It's a nice way to take yourself back to the beach from your home.

CORALS AND SEA PENS

Corals are cnidarians, in the same broad group as sea jellies. Most (but not all) hard corals are found in the tropics. The question 'is it an animal, vegetable or mineral?' is tricky to answer for corals – in some ways they could be considered all three, but are classed as animals. The animal part is the polyp – or rather, polyps, as corals are made up of hundreds or even thousands of of these tiny animals living together. Polyps trap food particles with their tentacles. As corals grow the polyps deposit minerals, building hard rock-like structures in which they live.

The varied colours of corals come from the algae, called zooxanthellae, which the polyps hold in their tissues – like gardens that they cultivate to get energy from the sun. When the water gets uncomfortably hot, corals expel their zooxanthellae to limit their physiological stress. At this point the coral loses its colour (a process called bleaching) but the polyps are still alive. If conditions improve rapidly enough they can start to regrow their algal gardens and recover; if the water stays warm for too long the coral colony will die. By the time broken pieces of coral skeletons wash up on the beach, both the polyp and the algae have normally gone.

OPPOSITE: Coral skeleton. Corals are diverse and can form a wide range of shapes, from branching, tree-like structures to compact spheres such as this one. The repeated flower-like indentations in the coral skeleton show where each polyp was positioned.

A 'sea pen' octocoral – octocorals are a group whose polyps have eightfold symmetry – washed up on a beach in New South Wales, Australia.
Photo Kerryn Wood

Sea pens are also colonies of polyps, but instead of all the polyps looking the same, one forms the central 'stalk' while others take on discrete roles, including feeding and reproduction. They don't build a hard skeleton. Sea pens anchor themselves in soft sediments and can resemble tall feathers. They are found from shallow waters to the deep sea.

SEA STARS

Sea stars and brittle stars are closely related and are all echinoderms (from the Greek for 'spiny skin').

A **brittle star** (ophiuroid) can be recognised by its discrete, pentagonal central disc from which its five arms radiate. These arms are long and often curl around on themselves, and can look hairy or spiky.

The central disc of a **sea star** (asteroid), on the other hand, is sometimes not clearly differentiated from its arms. Sea stars also have radial symmetry (like a wheel, with spokes) and don't always have five arms. They move using hundreds of 'tube feet' with suckers, which they operate through a hydraulic water-pumping system. When a sea star is ready for a meal, it everts its stomach out of its body to wrap it around the prey (for example, a snail). The food is then digested before the stomach is retracted. Sea stars can regenerate lost or damaged arms, and a whole animal can even regenerate from a single arm as long as part of the central disc remains.

Sea cucumbers (holothuroids) can also turn up on beaches. These often sausage-shaped echinoderms live on the sea floor. Many sea cucumber species are detritivores: they ingest sediment, digest the organic material and pass out 'clean' sand.

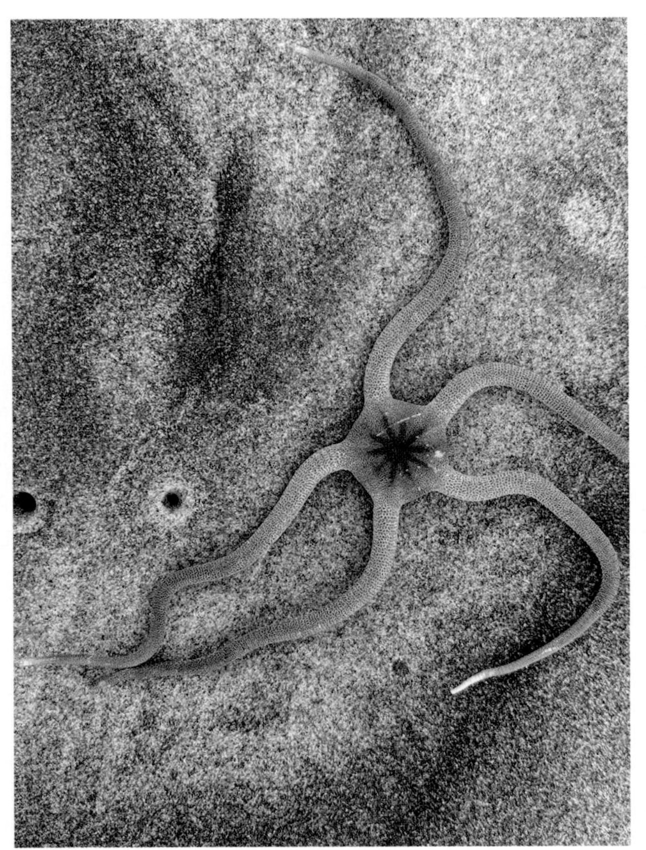

CLOCKWISE FROM TOP LEFT: Brittle stars have long slender arms and a distinct central disc.

A *Pentagonaster* or 'biscuit' sea star in New Zealand.

Sea stars use hundreds of hydraulically operated 'tube feet' to move around.

Sea cucumbers are the 'vacuum cleaners of the ocean', eating organic material from sediment.

Urchin tests rarely have spines attached when washed up on a beach, but you might find lone spines washed up on the sand as well.

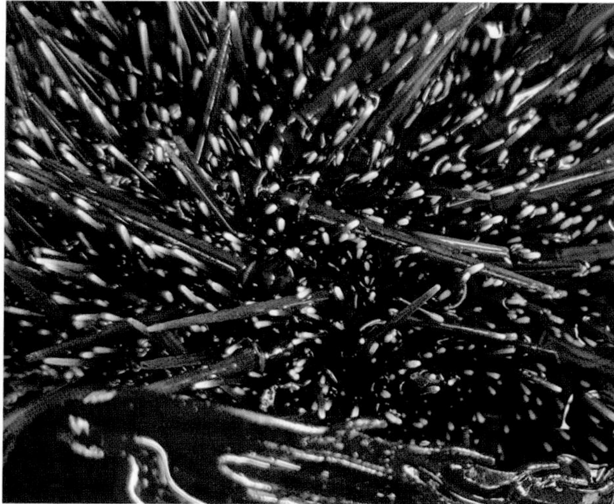

Live sea urchins have both primary (long) and secondary (shorter) spines, which they can move around.

URCHINS AND SAND DOLLARS

Sea urchins and sand dollars are related to sea stars and brittle stars – they're also echinoderms and have a five-spoked radial body plan. Urchins are like spiky balls crawling around the sea floor – they have a rounded test (their hard 'shell', made of calcium carbonate) and many spines. The spines are moveable, and although the urchin may use them to get around or dig into sediments, like sea stars they mainly move from place to place using hundreds of hydraulic tube feet. Their spines are primarily for protection, and in some species are venomous. When you find an urchin washed up on a beach it's not uncommon for the spines to be partially or completely absent, as they break off easily when the animal dies.

Sea urchins mostly eat algae. Some eat large brown seaweeds and, if populations grow too large, can decimate kelp forests. In the past few decades the black sea urchin *Centrostephanus rodgersii* has capitalised on warming waters to expand its range southward in eastern Australia, crossing Bass Strait to reach the delicious giant kelp forests of Tasmania. These ecologically important kelp forests have been hit hard

The sand dollar, a flattish urchin that burrows into sand.

'Sea hearts' are urchins whose shape is somewhere between the flat sand dollar and the ball-like sea urchin.

by the urchins, in many places disappearing entirely to be replaced by 'urchin barrens', which are like underwater deserts. In a balanced ecosystem, urchins are important herbivores, and are eaten by sea stars, lobsters and other predators. Overfishing of urchin predators – such as rock lobsters in Australia – can contribute to urchin population explosions.

The roe (the internal reproductive organs or gonads) of urchins is a valued delicacy. New Zealand Māori were harvesting *Evechinus chloroticus*, or kina, long before the arrival of Europeans; and in Australia two common species, the purple and red sea urchins *Heliocidaris erythrogramma* and *Heliocidaris tuberculata*, are commercially harvested.

Sand dollars (sometimes called sea biscuits or sand cakes – or pansy shells in South Africa) are urchins with flattened tests. When alive, their tests are hidden by fine spines covered with cilia (tiny hair-like structures), giving them a velvety feel. They live on sediments and often use their spines to burrow a little under the sand. They eat detritus or tiny algae and animals. Their spines are so delicate that you are unlikely to find the washed-up tests of these animals with many spines still attached.

SHARK EGGS

Sometimes on the beach you can find odd-looking brown, tough, leathery sacs: chondrichthyan egg cases (also known as mermaids' purses).

The Chondrichthyes – fishes whose skeletons are made of cartilage rather than hard bones – include sharks, skates and rays, and are a diverse group with a variety of ways of producing young. Some are viviparous, meaning the embryo develops inside the mother and is nourished by her; some are ovoviviparous, meaning eggs are produced, fertilised and hatch inside the mother; and some are oviparous, meaning eggs are laid and hatch outside the mother's body.

Shark and **skate egg cases** have evolved shapes and colours that help them to blend in with their environment. Oviparous shark egg cases are often attached to seaweed or rocks.The Port Jackson shark, a bottom-feeder horn shark that is common around southeastern Australia, lays a corkscrew-like egg case, the flanges of which help the case to stay wedged among rocks on the sea floor. Other horn sharks also lay spiralled eggs. Some shark and skate eggs look more like flattish, rectangular purses, bulging in the middle. Most have long tendrils that help the egg to become entangled in seaweed, reducing the chance that it will wash out to sea or onto a beach.

Most egg capsules only have one embryo, but some skate cases can have several. If you find an egg case, hold it up to the light to see if it has anything inside; if it's empty, there might be a small hole where the animal has already exited the capsule.

Because shark and skate egg capsules tend to dry out and become hard and a little shrunken on the beach, to fully appreciate their beauty you might like to rehydrate them by soaking them in water for a couple of hours.

Chondrichthyan embryos take several months to develop and emerge from the egg case: those of the Port Jackson shark take around 10–11 months. The vast majority of embryos do not make it – they're eaten from the egg case by predators such as other fish or turban snails. If a Port Jackson shark pup manages to develop fully and leave the egg capsule, it then has a reasonably good chance of long-term survival. Female Port Jackson sharks start to mate and lay eggs when they are 11–14 years old. Elephant fish – chondrichthyans with a trunk-like snout found around New Zealand coasts – mature a little more rapidly: the embryos develop in six to eight months, and females begin to reproduce when they are around five years old.

Many of the sharks and skates that lay the egg capsules you might find on the beach are not dangerous to humans: they are bottom-feeders, with teeth that resemble

Port Jackson shark egg cases have a spiral form that helps them become wedged between rocks, and tendrils to entangle in seaweeds.

plates of very rough sandpaper. They use these plates to crush and grind their prey, including molluscs (such as snails) and crabs. Many of the more pelagic open-ocean hunters, such as great white and tiger sharks, give birth to live young rather than laying eggs.

Shark and skate egg capsules come in a variety of sizes and shapes.

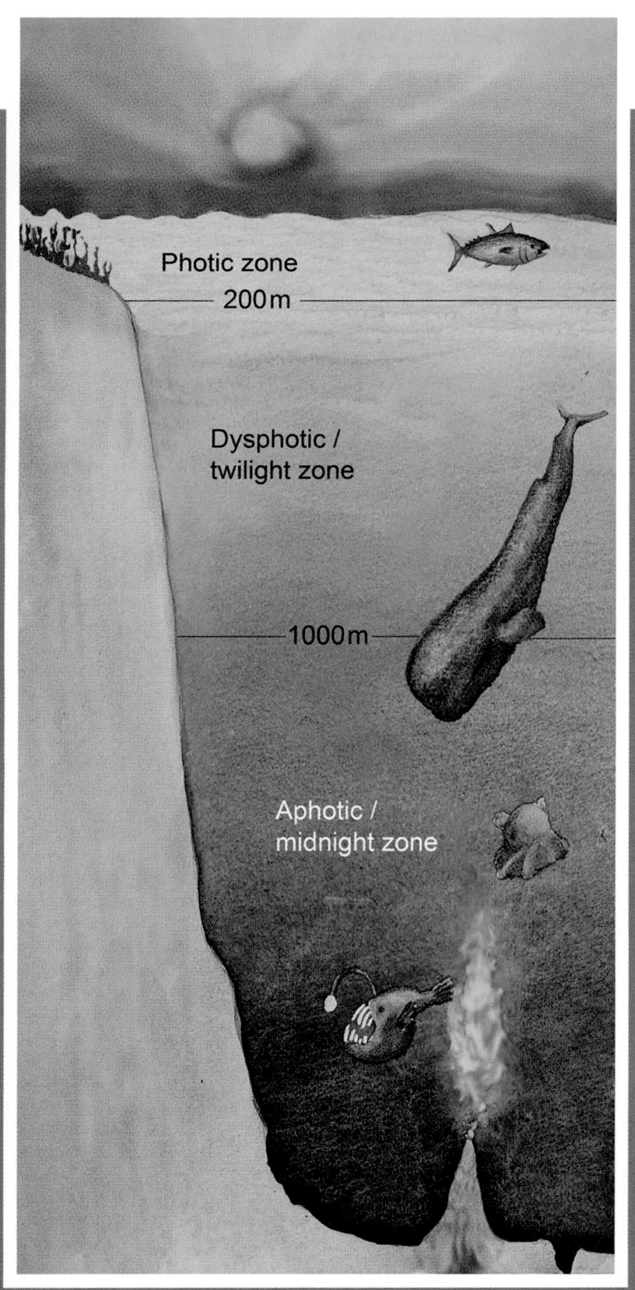

Some animals live hundreds or thousands of metres below the surface of the ocean, beyond the reach of sunlight.

Deep-water fish sometimes turn up on beaches, particularly near ports where fishing vessels discard bycatch before coming ashore.
Photo Nick Beckwith

Mycotophid 'laternfish' live about 1000m below the surface in the day, and have rows of photophores that light up along their sides. At night they migrate to the surface. *Photo Chris Woods*

From the Depths

Animals that live deep in the ocean beyond the reach of sunlight have evolved bizarre forms and lifestyles. Although most of them still get their energy indirectly from sunlight via organic particles that drift down from above, some have found ways to harness chemical energy from hydrothermal vents, where mineral-rich hot water issues from volcanically active plate margins.

The deeper you go underwater the bluer the light becomes, as longer wavelengths (the red end of the spectrum) are filtered out. Most light does not reach more than a couple of hundred metres down, beyond which is the twilight (dysphotic) zone – where the light is blue and dim. Because no photosynthesis can occur in the twilight zone, seaweeds do not grow. Beyond about 1000m is the midnight (aphotic) zone, where no light penetrates. The midnight zone is the realm of animals that can seem alien-like: fish with disproportionately big, sharp-toothed mouths and huge, powerful eyes capable of spotting the tiniest glimmers; glowing worms; giant crabs metres across; and octopuses with webbed tentacles.

Many animals in the deep sea use bioluminescence for hunting, mating or evading predators. They create light through carefully controlled chemical reactions, and can switch their light on and off and modify its intensity. The female anglerfish has an appendage dangling from her head, a glowing lure that captivates and entices prey to come close to her huge, toothy mouth.

Very occasionally, dead deep-sea animals find their way onto beaches. These have normally been brought to the surface by humans and discarded overboard as bycatch, to sink or be eaten. Some fish have buoyancy control devices called swim bladders that regulate their vertical position in the water. When these animals are brought to the surface suddenly their bladder over-inflates, often killing the animal and making it unlikely that the fish will sink back into the depths. Even those without swim bladders are likely to die when brought rapidly to the surface: the release of water pressure causes barotrauma, where liquids and gases expand rapidly and rupture tissues.

A skua is silhouetted against the sun's rays as the icebreaker SA *Agulhas* passes Marion Island in the sub-Antarctic Indian Ocean, 2007.

Seabirds

Dead seabirds are common on beaches, and can sometime appear in surprisingly large numbers, with hundreds or thousands of dead or dying birds in the same place at the same time. Several factors can lead to these mass mortality events, including food shortages, storms and toxins.

Many seabirds migrate, sometimes in vast flocks. The **short-tailed shearwater**, for example, which nests in burrows in southern Australia, flies to the Northern Hemisphere every year, setting off in autumn when its chicks are large enough to fend for themselves. Tens of thousands of shearwaters migrate together. On their journey to the Arctic they cover some 15,000km over about six weeks. To survive this migration they must be in peak condition and have access to good food (mostly krill and other pelagic crustaceans) along the way. The **sooty shearwater** (tītī, or muttonbird), which breeds in Australia, New Zealand, Chile and the Falkland Islands, has a similar pattern of breeding and migration.

With warming waters and overfishing, however, many birds are finding it harder to feed during migration, and mass starvation can occur. Storms also kill seabirds, or

Shearwaters sometimes die during foraging or migration. *Photo Nick Beckwith*

Remains of a dead seagull. *Photo Nick Beckwith*

CLOCKWISE FROM ABOVE: A wandering albatross and chick on a raised nest mound, Marion Island, 2007. Albatross chicks are often left alone on the nest while parents feed at sea and are vulnerable to attack by introduced predators, including mice. Adult wandering albatrosses travel thousands of kilometres.

Dead penguins found with only foot and flipper flesh intact have probably fallen prey to predators such as seals.

King penguins on Marion Island.

strong winds exhaust them while they're foraging or migrating. Toxins are another cause of mass mortality. For example, red tide blooms of algae sometimes produce toxins which accumulate in the prey of birds and make them sick. Red tides also produce surfactants that cause seabird feathers to become less water-resistant. Without this protection, as a result of direct exposure to cold water a bird can die from hypothermia.

Starvation, illness or injury can also cause more solitary seabirds, such as penguins or albatross, to die and wash up on beaches, and death as a result of ingesting plastics is an increasing problem. A piece of plastic drifting in the ocean can look a lot like a delicious sea jelly or crustacean, and soft plastics such as balloon fragments seem to be particularly fatal. One study from South America showed that up to 35 percent of **Magellanic penguins** found dead on beaches had plastic in their stomachs. Seabirds – especially scavenging seagulls – are also prone to entanglement from string, fishing line and plastic rings.

A moulting rockhopper penguin on Marion Island.

Penguins nest along coastlines all around the Southern Hemisphere, including in quite hot areas of Africa and Australia: the **little blue penguin** as far north as Coffs Harbour on the east coast of Australia, around latitude 30°S, and the **African penguin** as far north as Hollams Bird Island off Namibia, around 24°S. Penguins go through an annual moult for several weeks, during which time they lose and regrow their feathers. While penguins are moulting their feathers are not waterproof, and to avoid hypothermia they must stay out of the water. Penguins occasionally begin their moult when they are far from home, and these vagrant birds might be seen on beaches well out of their normal breeding range. **King penguins**, for example, sometimes moult in New Zealand, although no breeding colonies of this species exist there.

ABOVE: An elephant seal rests on a pile of giant kelp, Falkland Islands.

OPPOSITE TOP RIGHT: Sea lions are increasingly common on beaches around New Zealand, particularly in the South Island.

OPPOSITE BOTTOM RIGHT: Sea lions and seals often come onto beaches or rock platforms to rest. This sea lion is cunningly masquerading as a log, but its disguise is foiled by its tracks.

Here Be Giants

MARINE MAMMALS

Whales and **dolphins** are mammals, not fish: they evolved from land mammals that 'returned' to the sea tens of millions of years ago, and they breathe air, regularly surfacing for this purpose (or, as Lewis Carroll might have said, 'for this porpoise'). **Seals** and **sea lions** are also marine mammals but are able to spend time both on land and in the sea, coming up onto beaches, rock platforms and into dune vegetation to rest or breed. If you see a seal or sea lion on a beach, please keep your distance.

ABOVE: Elephant seals come ashore to moult.

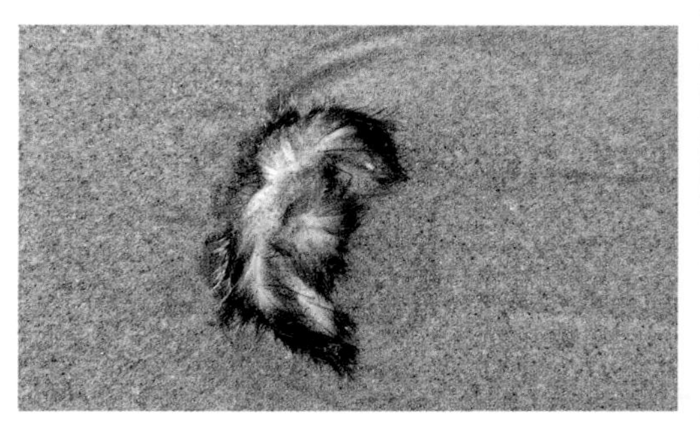

A piece of seal fur washed up on a beach in Otago, New Zealand. Many such patches of pelt were found on a day when killer whales were active just offshore.

A whale vertebra.

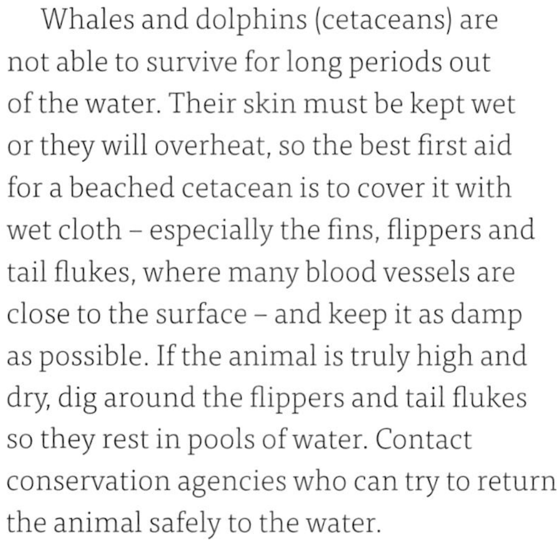

Whales and dolphins (cetaceans) are not able to survive for long periods out of the water. Their skin must be kept wet or they will overheat, so the best first aid for a beached cetacean is to cover it with wet cloth – especially the fins, flippers and tail flukes, where many blood vessels are close to the surface – and keep it as damp as possible. If the animal is truly high and dry, dig around the flippers and tail flukes so they rest in pools of water. Contact conservation agencies who can try to return the animal safely to the water.

Nearly all whale and dolphin species have been known to strand, either singly or en masse. Those that strand alone are usually sick or injured in some way, and autopsies sometimes show high levels of pollutants that cause organ damage and compromise the immune system; others have plastic in their stomachs.

Mass strandings happen only in the most social toothed whales. In some of these species the members of a pod – a group of whales – are closely related to one another. The first to strand is often ill or may have just given birth. Because of social cohesion within the pod, the others tend to stay close by and can also become beached as the tide recedes. These deep-water species might not realise the risk that shallows pose, especially in places with large tides and gently sloping beaches.

Although social bonds can contribute to mass strandings, they also help with rescue efforts. Keeping the group together as much as possible and organising rescue attempts around the most socially important members (such as females with calves) sometimes helps to get the pod to head back into safe water.

CLOCKWISE FROM TOP LEFT: Dead sperm whale with injuries, New Zealand. *Photo William Rayment*

Stranded and dead humpback whale, southeastern Australia. *Photo Kerryn Wood*

Stranded Gray's beaked whale, Birdlings Flat, New Zealand. *Photo Steve Dawson*

Jack Aplin with a giant squid, found on a New Zealand beach.
Photo Dan Aplin

GIANT SQUID

Occasionally, huge creatures from the deep appear on a beach, such as the **giant squid** *Architeuthis*. The average length of a giant squid (from records of found specimens) is 11m from the tip of the mantle to the far ends of the tentacles and its eyes can be 30cm in diameter. Giant squid live in the deep sea, usually below about 300m and sometimes at more than 1000m. They feed on fish and other cephalopods, including other giant squid.

We don't know a great deal about these creatures as their habitat makes them difficult to study, but we do know that they are eaten by **sperm whales**, as giant squid beaks are sometimes found in guts of these great leviathans (one stranded whale was found to have 47 beaks in its stomach!) and some whales bear the sucker marks of ferocious battles with giant squid. Giant squid are thought to live for a few years only (a maximum of 14, but probably fewer), meaning their growth rate must be remarkable – perhaps up to around 5mm per day. They have two feeding tentacles armed with sharp suckers ringed with teeth, and can shoot these out to catch prey from several metres away. Like most cephalopods, they also have eight arms. They move by jet propulsion.

RIGHT: A dead sunfish found in Otago Harbour, New Zealand. *Photo Linda Groenewegen*

SUNFISH

Sunfish are crazy-looking animals and are enormous and heavy – up to 1000kg and, in rare cases, over 2000kg. They have a tall fin on the top and bottom and an oddly truncated rear end, so that their gigantic body seems to end abruptly with a stumpy tail.

Sunfish look like they should be bad at swimming – and they are fairly awkward. They use their stumpy tail more as a rudder than for propulsion and their caudal and anal (top and bottom) fins a little like oars. They can move quickly when they want to, however, and may leap right out of the water (breaching) to help remove some of their many parasites.

There are several sunfish species, and not all grow to be enormous. *Mola mola* is perhaps the best-known and biggest. Sunfish are not dangerous to people – they mainly feed on small animals such as fish, squid, sea jellies and plankton. The female can produce hundreds of millions of eggs at a time; her babies look a little like pufferfish, with star-like spines.

ABOVE: The spinifex seedhead is a large spiky ball, well suited to dispersal by wind and water.

LEFT: Mangrove and spinifex propagules washed up at the high-tide mark on a beach near Auckland, New Zealand.

Southern Connections

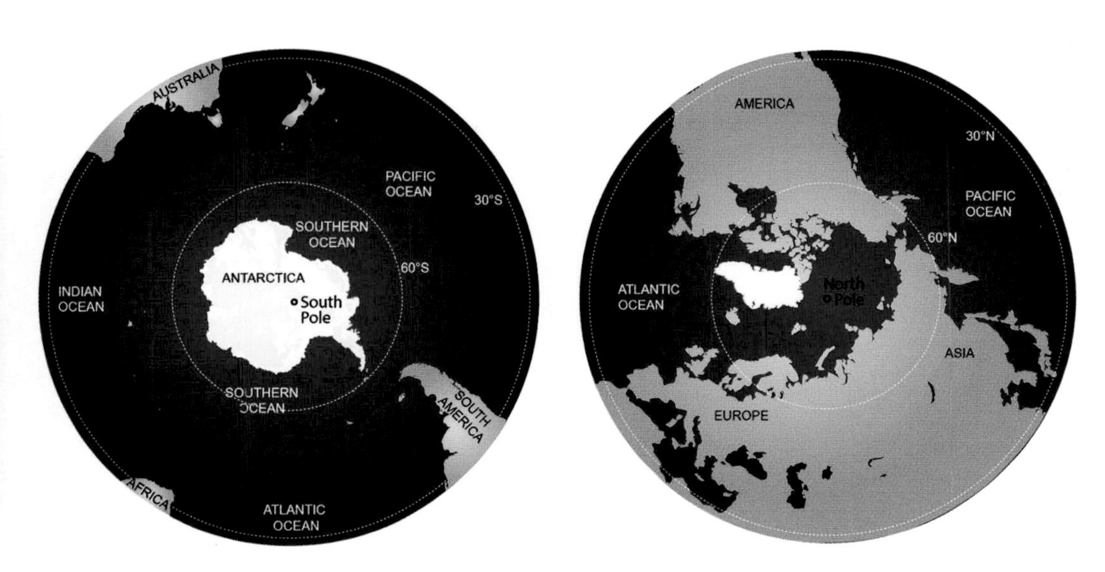

These polar-centric views of parts of the Southern and Northern hemispheres show that the ratio of land to sea is greater in the north.

Unlike the Northern Hemisphere, where continents stretch almost to the pole, the Southern Hemisphere has relatively little land and a lot of sea – about 80 percent of the Southern Hemisphere is ocean compared to about 60 percent of the Northern Hemisphere. The Southern Ocean encircles Antarctica and connects all the major ocean basins: the Atlantic, Indian and Pacific. The Southern Ocean is also home to the strongest ocean current in the world, the Antarctic Circumpolar Current, which sweeps eastward around the globe and is a bit like an oceanic roundabout, carrying water and anything it holds from place to place. Southern Hemisphere beaches are connected by this and other strong ocean currents.

PLANT SEEDS

Terrestrial plants have developed clever strategies for getting their seeds to new places. Many rely on the wind for aerial dispersal of spores or seeds, but some, especially those that grow near the coast, have evolved buoyant propagules (seeds or other parts that can grow into a new plant) that can float some distance at sea.

Sea beans are the hard, buoyant seeds of several tropical plants. Because they come from different species, they have a range of sizes and shapes. Some can stay buoyant for years, and many have smooth exteriors, perhaps to reduce drag at sea.

Coconuts, the fruit of the tropical *Cocos* palm, are adapted for oceanic dispersal. They float well, can drift at sea for several weeks, are salt-tolerant and grow in sandy soils, so can germinate close to the beach on making landfall. Coconut palms are found in many places, but whether their wide distribution is the result of natural dispersal or human activity has long been a subject of scientific debate. As with most such cases, the answer is probably a bit of both. Although humans have surely helped to carry and cultivate coconuts, dispersal of some of the fruits predates human migrations: fossil coconuts from millions of years ago have been found in both New Zealand and India.

Spinifex grasses grow on beach dunes around the Southern Hemisphere, including in Africa, Asia, New Zealand, Australia and the Pacific Islands. Their seedhead is a pom-pom of spikes that can be bowled along the beach by the wind and float for some distance at sea. Sand dunes are easily eroded by water or wind, and grasses such as these play important roles in stabilising dunes and reducing coastal erosion. Try to avoid stepping on dune plants at busy beaches as excessive foot traffic will destroy them and speed up erosion.

A mangrove propagule from the New Zealand native *Avicennia marina* subspecies *australasica*, washed up on a beach near Auckland, begins to put out roots.

Mangroves also play a critical role in mitigating coastal erosion, along with providing a range of other ecosystem services. They are found mostly in tropical and sub-tropical regions, in areas that are inundated at high tide. Because mangroves usually grow in silty, muddy soils that are low in oxygen, some have evolved special roots that send up 'snorkels' – pneumatophores – that carry oxygen down to the roots. These pneumatophores look like the bristles of huge hairbrushes poking out of the mud. As mangroves grow right on the edge of the ocean, oceanic dispersal is a perfect way of getting to new areas. Fertilised flowers produce propagules a bit like seeds, and these can float for a few days and have a good supply of nourishment to help the young plant get started in life once it makes landfall.

PUMICE

When a volcano erupts under the ocean the molten lava mixes with water and gases and cools quickly, forming frothy rock with holes where the gas has escaped. Because many of the holes initially contain air, this rock – **pumice** – can float. Eventually the holes fill with water and the pumice will sink, but some pieces stay buoyant for months and travel long distances at sea.

In 1962 a volcano erupted in the South Sandwich Islands between South America and the Antarctic Peninsula, spewing pumice into the sea. Rafts of this rock set sail in the Southern Ocean and were recorded in far-distant places over the following two years. Some even reached Tasmania and New Zealand around 18 months later, having travelled eastward more than 13,000km in the Antarctic Circumpolar Current.

Likewise, in 2012 a submarine volcano erupted in the southwest Pacific in the Kermadec chain. The pumice rafts spread out, most heading west, and washed up in great quantities on eastern Australian beaches over several months. Many pieces were covered in living plants and animals including algae, barnacles and corals. Scientists think rafting of this kind plays a critical role in connecting ecosystems, and could even 'seed' new coral reefs after destruction of older reefs, or when new habitats open up, for instance around new volcanic islands.

Although volcanoes create pumice in all of the world's oceans, this buoyant rock

Pumice from the 2012 eruption in the Kermadecs washed up on beaches in southeastern Australia in 2013–14. Many were encrusted with living algae, bryozoans and other animals. This specimen includes bryozoans (b), serpulid polychaete worm tubes (p), anemones (a) and goose barnacles (g). *Photo Alexander Harrison*

Pumice on a beach in southern New South Wales. *Photo Kerryn Wood*

Some bryozoans form a crust on rocks and other surfaces.

Bryozoans and hydroids are animals that are almost exclusively colonial and often attach to other organisms. Bryozoans are filter feeders, meaning they sieve nutrients from the water. Colonies include both feeding and non-feeding individuals. Here colonies of a soft bryozoan (red) and hydroids (brown, feather-shaped) are attached to beach-cast kelp. *Photo Mia Ching*

is particularly commonly produced in the Pacific Ocean. If you find pumice on a beach, examine it closely. If it is clean, it has either been tumble-cleaned by waves on the sand or has spent very little time at sea and might have come from a nearby eruption. If it has a crusting of algae, bryozoans (which look like lace but are really colonies of invertebrates) or other animals, those creatures might have travelled a long way, rafting with the floating rock.

Pumice has long been valued by humans for its abrasive qualities: its rough surfaces are great for rasping dead skin from calloused feet!

KELP RAFTING

In his poem 'Seaweed', nineteenth-century poet Henry Wadsworth Longfellow mused:

> *Ever drifting, drifting, drifting*
> *On the shifting*
> *Currents of the restless main;*
> *Till in sheltered coves, and reaches*
> *Of sandy beaches,*
> *All have found repose again.*

When robust and buoyant seaweed becomes detached, for example during a storm, it can drift remarkably long distances. Although most species can't reattach when they hit the coast again, many are still able to reproduce even after a long time at sea – particularly in colder waters, where the kelp tissue degrades more slowly.

People have long speculated on whether drifting seaweed could help to carry plants and other animals across oceans. In the 1950s a large jack rabbit was found drifting on a *Macrocystis* kelp raft more than 60km from the Californian coast. Recently, scientists have been using genetic tests to work out just how far seaweed can travel at sea. Because DNA is different not just among species but also between individuals of the same species (your DNA is different to your friend's DNA – and even, unless you are an identical twin, a bit different to that of your brother or sister), we can use it to work out how individuals are

Buoyant southern bull kelp, *Durvillaea*, washed up on beaches in New Zealand.

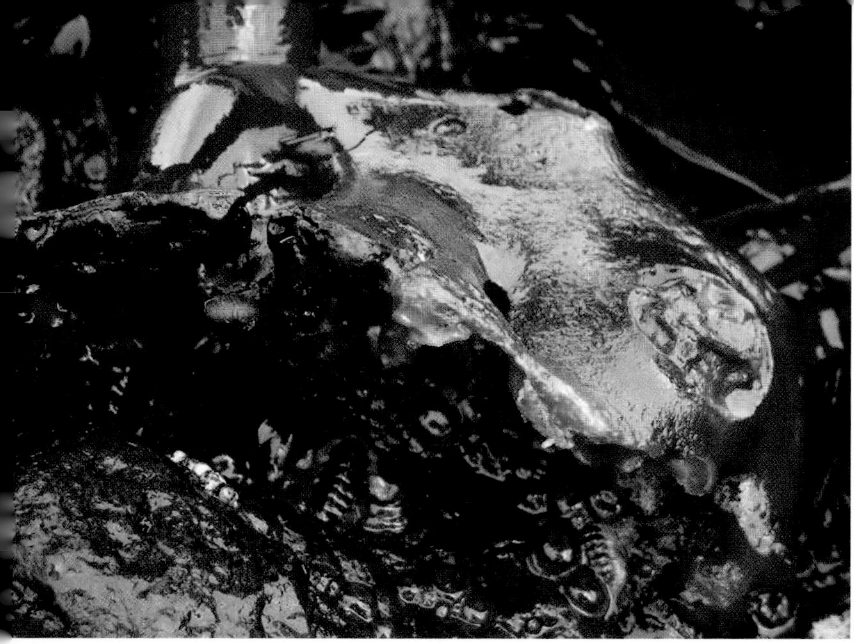

LEFT: Bull kelp holdfasts are hollowed out by burrowing animals and can host many different species, which can live for generation after generation in the same holdfast.

MIDDLE LEFT AND BOTTOM: *Durvillaea* bull kelp from the sub-Antarctic found on St Clair Beach in 2009. Note the large goose barnacles. These kelp rafts had many animals living in their holdfasts.

related to each other. In many cases, that includes being able to work out from which population an individual has come.

In 2009 following stormy southeasterlies, a lot of southern bull kelp (*Durvillaea antarctica*) washed up on St Clair Beach in southern New Zealand. Although the same species grows on rock platforms nearby, the large goose barnacles attached to the beach-cast kelp indicated that it had been at sea for a long time. A number of rafts also had holdfasts attached that were packed full of small animals, including crustaceans, echinoderms (such as sea stars) and molluscs. Genetic tests showed the DNA of the kelp and the animals in its holdfasts matched that of populations in the sub-Antarctic Snares and Auckland Islands, 400–600km away.

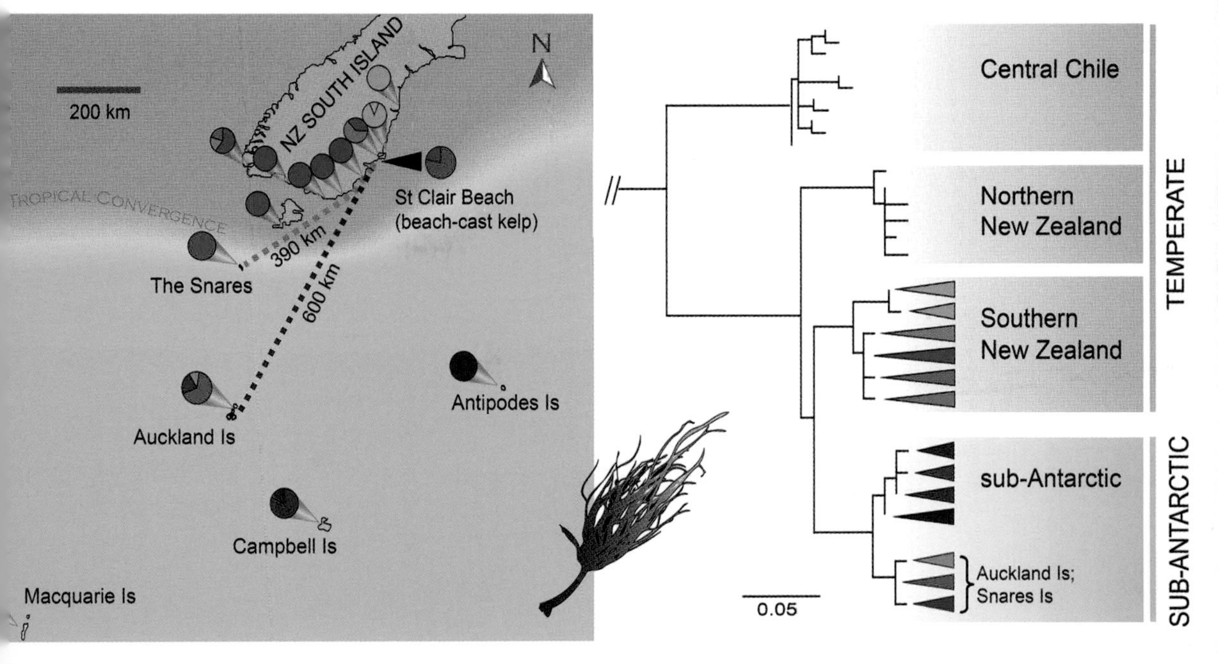

DNA tests revealed that the kelp rafts found on St Clair Beach and the animals in the holdfasts closely matched populations in the sub-Antarctic Auckland and Snares islands. Left: Coloured pie charts show the genetic lineages of bull kelp found around the southern New Zealand region. Right: Phylogenetic 'trees' are similar to family trees. This one shows how genetically similar the coloured lineages shown on the map (left) are to each other. Each branch tip along the left is a different lineage, and the horizontal distance along branches between two lineages represents how genetically different they are. These analyses show the kelp rafts had drifted hundreds of kilometres at sea. Figure modified from Fraser et al. 2011. See Further Reading.

The St Clair study was exciting because it showed that kelp really could carry animals long distances at sea. Just how far kelp rafts might travel, though, was still not clear. Genetic evidence from bull kelp (*Durvillaea*) growing around the sub-Antarctic islands – a ring of small isolated islands in the vast Southern Ocean – suggested that most of these kelp populations, and the animals in their holdfasts, had only recently colonised these islands.

In fact, scientists think the islands were colonised by kelp after the last ice age, which reached its coldest point some 18,000–20,000 years ago. Expanding sea ice reached across the Southern Ocean, and probably scoured all the kelp communities from the sub-Antarctic islands within its reach. The genetic similarities among these kelp communities suggest that the recolonisation of the island happened from the same source region. Scientists

Kelp attaches so firmly to the rock surface that often the rock will break before the kelp does!

speculate that as the climate warmed and the sea ice started to retreat back towards Antarctica, kelp that broke off from somewhere further north was able to form rafts that seeded new populations on the islands as they were freed by the receding ice. The same patterns have been found for giant kelp, *Macrocystis pyrifera*, around the sub-Antarctic, and animals associated with seaweeds.

But could kelp rafts really travel thousands of kilometres between sub-Antarctic islands, or cross the vast Pacific Ocean between New Zealand and Chile? In 2017 a couple of pieces of *Durvillaea antarctica* were found on beaches on King George Island, near the Antarctic Peninsula. This find was surprising, because bull kelp doesn't grow in the Antarctic region, and scientists thought that it was almost impossible for floating objects drifting at the surface to cross the Southern Ocean from the sub-Antarctic to reach Antarctica. New DNA analyses were able to show exactly where the kelp had come from: one piece had travelled more than 20,000km across the sea from the Kerguelen Islands; the other had come from South Georgia, and modelling suggests it travelled around 25,000km to reach King George Island.

Here at last was evidence that kelp rafts can travel thousands – even tens of thousands – of kilometres. And the only way they could have crossed all the barriers the Southern Ocean presents – especially the circumpolar frontal subduction zones at the boundaries of cooler southern, and warmer, more northern waters – would be with the help of wind and waves.

These 'taxis of the sea' must be a powerful force in helping to shape coastal ecosystem biodiversity in the Southern Hemisphere. As the climate warms and many plants and animals try to move towards the cooler poles to track suitable environments, mechanisms such as kelp rafting will be

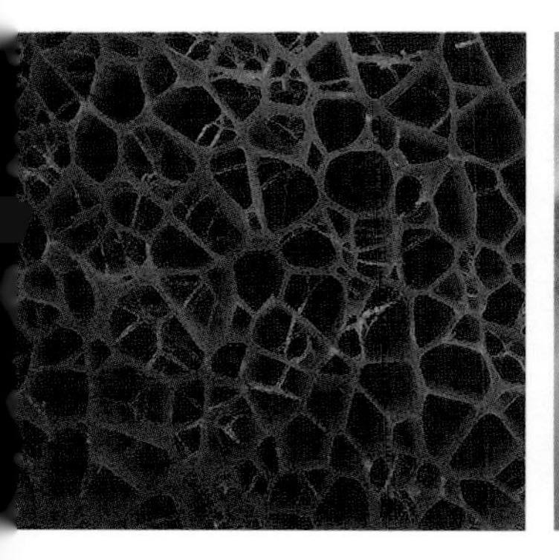
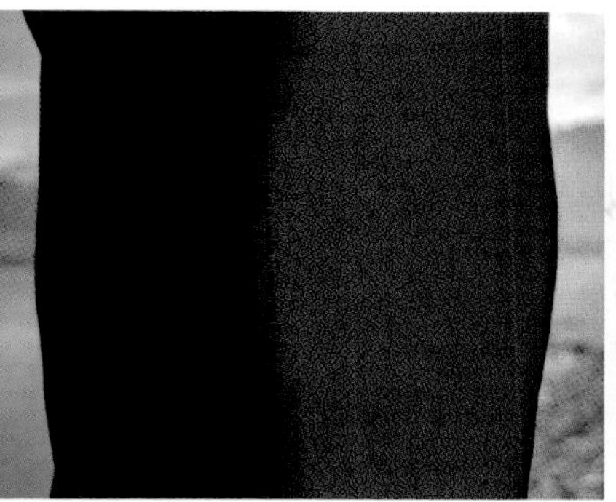

Some *Durvillaea* species float well because of the gas-filled honeycomb-like cells of their fronds.

Durvillaea bull kelp in Australia has dense, non-buoyant fronds and does not travel far. Any cast up on the beach is generally from nearby rock platforms or shallows.

important for the survival of some species. We tend to think of nature conservation as meaning that things should stay the same, but in a rapidly changing world, plants and animals (and algae) will need to move to new locations. Ecosystems will have to change to survive.

Driftwood can travel long distances at sea.

DRIFTWOOD

Storms often wash large logs and freshly torn-up trees into the sea. When fresh or dry, most wood can stay afloat for a long time; it may sink when it becomes saturated. Driftwood also makes an excellent raft and has been implicated in the movement of iguanas and insects across the seas.

Many species of southern beech trees (*Nothofagus*) occur in Australia, New Zealand, South America and New Guinea. They also used to grow in Antarctica before the ice took over, and have a long fossil record dating back to the time when all southern landmasses were joined in the supercontinent Gondwana. Genetic analyses show, however, that the current distributions of species can't be explained by the breakup of Gondwana alone. Instead, long-distance dispersal of *Nothofagus* must have occurred at some time. That means the trees somehow crossed oceans and survived. These days, *Nothofagus* trees sometimes wash up on beaches on the treeless Falkland Islands. They must drift across hundreds of kilometres of ocean, from South America or perhaps even New Zealand, to reach the islands.

The driftwood you find on the beach might have come from a nearby river mouth, but could also have travelled a long, long way.

The Future of Southern Beaches

Beaches are dynamic ecosystems that constantly shift and change shape. Lying at the interface of terrestrial and marine environments, beaches experience some of the most extreme conditions of both worlds – smashing waters advancing and receding twice a day, baking hot sunshine or freezing cold storms.

Slight modifications in coastal topography can alter coastal ocean dynamics and sometimes cause large-scale changes to beach erosion or sand deposition, carving coastlines into new shapes and occasionally eating into urban areas. The erection of breakwaters, piers and other artificial structures can also have a drastic impact on beach dynamics.

Nowadays beach ecosystems are facing even greater challenges. Global warming is causing ice sheets to melt and water to expand, leading to sea-level rise. Following the last ice age, around 20,000 years ago, sea levels rose well over 100m (the amount varied around the world – in many places it was about 120m). That change happened over millennia, giving most species time to shift with the rising waters. These days, however, the climate is changing with unusual speed. Hundreds of millions of people live in coastal areas that are less than 10m above sea level, and many major cities are built at the edge of seas and harbours. Predicted sea-level rise of tens of centimetres in the next few decades will have major impacts, both on coastal communities and coastal ecosystems.

Rapid population growth is also putting intense pressure on marine ecosystems. Food webs – the network of 'who eats who' – become unbalanced by changes at the top (to predators such as sharks and other fish) or the bottom (through damage to seagrass or algal 'primary producers'), and these imbalances send shock waves of change through ecosystems. Marine ecosystems are only just starting to recover from intensive whaling, sealing and penguin-oil exploitation, activities that removed many top predators in the 1800s and 1900s.

The challenge of mitigating our impacts on the ocean can seem overwhelming. From global warming to plastic pollution, humanity is wreaking havoc on marine ecosystems. But we're smart, and we have the tools to change things for the

A tripot, once used for boiling penguins for oil, left on a beach on Marion Island in the sub-Antarctic Indian Ocean.

CLOCKWISE FROM TOP LEFT:
A dead seahorse.

Beaches often hold clues to local history. Railway nails can still be found around beaches near ports that were serviced by rail more than 100 years ago.

A shark's jaw.

better. Simple actions, such as choosing renewable energy sources, buying food in biodegradable packaging (or no packaging at all), choosing certified sustainable seafood, or making sure you dispose of plastic rubbish carefully, have positive impacts.

Our beautiful southern beaches will once again glow with good health when the oceans they represent are clean and bursting with life.

Further Reading

Serolid isopods on a beach in southern Chile.
Photo Huw Griffiths

This section is a list of resources for those who want to know more about their finds on the beach. While the publications do not cover the full scope of material mentioned in this book, they are useful starting points.

There are also many useful online resources, but as these can change over time I have not included many websites in this list. When searching online for information, be wary: websites can contain errors and exaggerated or outdated information. Reliable and up-to-date information can usually be found on sites hosted by scientific agencies such as the National Oceanic and Atmospheric Administration (US), the National Institute of Water and Atmospheric Research Taihoro Nukurangi (NIWA, NZ), the Commonwealth Scientific and Industrial Research Organisation (CSIRO, Australia), or reputable museums and universities. NIWA, for example, has produced a range of interactive guides to marine invertebrates and seaweeds in New Zealand, which provide information as well as identification tools.

General

Collins Field Guide to the New Zealand Seashore by Sally Carson and Rod Morris (Auckland: HarperCollins, 2017), 416pp. (Interesting information and useful pictures to help you identify and learn more about nearshore marine life in New Zealand.)

Australian Marine Life by G. Edgar (Frenchs Forest: New Holland, 2012, 2nd edn.), 624pp. (A useful pictorial guide for identifying many common shallow-water marine plants and animals. Although it was written for Australia, some of the species and many of the genera are common throughout the Southern Hemisphere.)

'The ecology of rafting in the marine environment. II: The rafting organisms and community', by M. Thiel and L. Gutow, in ***Oceanography and Marine Biology: An annual review*** 43, (2005), pp. 279–418. (A scientific review paper that summarised the types of creatures that have been observed or inferred to raft at sea.)

Ambergris

Floating Gold: The search for ambergris, the most elusive natural substance in the world by Christopher Kemp (Auckland: HarperCollins, 2012), 320pp. (This book was written for a general audience and covers the origin of and search for ambergris, with a focus on the author's New Zealand experiences.)

Bioluminescence

'Bioluminescence in the sea', by S.H.D. Haddock, M.A. Moline and J.F. Case, in ***Annual Review of Marine Science*** 2 (2010), pp. 443–93: www.annualreviews.org/doi/10.1146/annurev-marine-120308-081028. (A scientific review paper that summarised what we know about bioluminescence, in various forms, in the oceans.)

Giant squid

'Unanswered questions about the Giant Squid *Architeuthis* (Architeuthidae) illustrate our incomplete knowledge of coleoid cephalopods', by C.F.E. Roper and E.K. Shea, in ***American Malacological Bulletin*** 31, 1 (2013), pp. 109–22: https://doi.org/10.4003/006.031.0104. (A scientific paper summarising what we know – and don't know – about giant squid.)

Goose barnacles

'Rates of growth in two species of *Lepas* (Cirripedia)', by T.M. Skerman, in ***New Zealand Journal of Science*** 1, (1958), pp. 402–11. (A scientific paper that recorded the growth rates of goose barnacles that settle on objects drifting at sea. The information can be used to estimate the minimum time that an object has been drifting.)

Indigenous uses of coastal resources

'Marine conservation: Māori and the sea', by Gerard Hutching and Carl Walrond, Te Ara – the Encyclopedia of New Zealand: www.TeAra.govt.nz/en/marine-conservation/page-1

'Sea Country: An Indigenous perspective', ***The South-East Regional Marine Plan Assessment***, (Hobart: National Oceans Office, 2002), 186pp.

Plastics

'Plastic debris in the ocean: The characterization of marine plastics and their environmental impacts, situation analysis report', by F. Thevenon, C. Carroll and J. Sousa (eds) (Gland, Switzerland: IUCN, 2014): https://portals.iucn.org/library/node/44966

'Primary microplastics in the oceans: A global evaluation of sources', by J. Boucher and D. Friot (Gland, Switzerland: IUCN, 2017): https://portals.iucn.org/library/node/46622

(These two International Union for Conservation of Nature reports provide summaries of the scale of the problem of plastics in the oceans.)

Seaweeds and seaweed rafting

New Zealand Seaweeds: An illustrated guide by Wendy Nelson (Wellington: Te Papa Press, 2020), 352pp. (A photo-rich guide for identifying common seaweed species. Although written for New Zealand, some of the species/genera are found throughout the Southern Hemisphere.)

'Oceanic rafting by a coastal community' by C.I. Fraser, R. Nikula and J.M. Waters, in ***Proceedings of the Royal Society B*** 278 (2011) pp. 649–55: http://dx.doi.org/10.1098/rspb.2010.1117. (This scientific paper used DNA analyses to work out where beach-cast bull kelp – and associated animals – found in New Zealand had come from.)

'Kelp genes reveal effects of subantarctic sea ice during the Last Glacial Maximum' by C.I. Fraser, R. Nikula, H.G. Spencer and J.M. Waters, in ***Proceedings of the National Academy of Sciences of the United States of America*** 106 (2009), pp. 3249–53: http://dx.doi.org/10.1073/pnas.0810635106.) This scientific paper used DNA analyses to infer that bull kelp had recolonised the sub-Antarctic islands since the last ice age.)

'Antarctica's ecological isolation will be broken by storm-driven dispersal and warming' by C.I. Fraser, A.K. Morrison, A.M. Hogg, E.C. Macaya, E. van Sebille, P.G. Ryan, A. Padovan, C. Jack, N. Valdivia and J.M. Waters, in ***Nature Climate Change*** 8 (2018), pp. 704–08: http://dx.doi.org/10.1038/s41558-018-0209-7. (A scientific paper that used genomic (DNA) and modelling analyses to work out where bull kelp found in Antarctica had come from, and how it got there.)

'The biogeographic importance of buoyancy in macroalgae: A case study of the southern bull-kelp genus *Durvillaea* (Phaeophyceae), including descriptions of two new species' by C.I. Fraser, M. Velásquez, W.A. Nelson, E.C. Macaya and C.H. Hay, in ***Journal of Phycology*** 56 (Oct. 2019), pp. 23–36. (A scientific paper summarising the known species of southern bull kelp. The figure reproduced on pp. 52–53 is by Ceridwen Fraser and is used under licence agreement with the copyright holder, the ***Journal of Phycology***, Wiley, 2020.)

'Rafting of jack rabbit on kelp' by J.H. Prescott, in ***Journal of Mammalogy*** 40 (1959), pp. 443–44. (A scientific paper recording the sighting of a jack rabbit on a kelp raft about 24km from land.)

Southern beech tree dispersal

'Relaxed molecular clock provides evidence for long-distance dispersal of *Nothofagus* (southern beech)' by M. Knapp, K. Stöckler, D. Havell, F. Delsuc, F. Sebastiani and P.J. Lockhart, PLoS Biology 3 (2005), e14: http://dx.doi.org/10.1371/journal.pbio.0030014. (A scientific paper that used genetic 'molecular clock' analyses [using the number of DNA mutations between populations, and an estimate of the rate mutations occurred] to infer that southern beech trees must have achieved long-distance, trans-oceanic dispersal.)

Acknowledgements

I am enormously grateful to Rachel Scott, Imogen Coxhead, Fiona Moffat and the editorial board of Otago University Press for their support and encouragement. This book became my main 'lockdown project' during the Covid-19 crisis, helping to keep me from feeling overwhelmed by the horrors unfolding globally, and allowing me to focus instead on the fascinating and wonderful stories that the ocean can tell.

A huge thank you to Keith Probert and three anonymous reviewers, who read and gave feedback on the full draft manuscript. I am also grateful to the other scientists who were kind enough to provide input and feedback on some parts of the book, particularly Dave Craw, Steve Dawson, John Bolton, Shelley Dixon, Lloyd Esler, Linda Groenewegen, Alexander Harrison, Pat Hutchings, John Jillett, Miles Lamare, Janice Lord, Georgina Pickerell and Ata Suanda. Pamela Olmedo Rojas kindly supplied the recipe for kelp stew. Thanks to the Tasmanian Aboriginal Centre for providing *palawa kani* beach words and permission to print them in this book, and additional audiovisual resources for the online educational material. Thanks to Lawrence Perry from the Wonnarua National Aboriginal Corporation for checking and approving use of Wonnarua words. Thanks to Sonia Verónica Millahual Cheuque – the lonko (leader) and lawentuchefe (doctor) of the Francisco Trecan Indigenous Community of Los Pinos Island, Chile – for translating words into Mapuche, and to Pamela Olmedo Rojas and Alexandra Gangas for translation assistance. Thanks to members of the East Otago Taiāpure for checking the Māori words selected for the language table. Several people contributed photos for the book, for which I am grateful. Thanks, too, to my husband Amit, for helping to keep our son entertained while I was writing, and to my parents, for their eternal support. I also thank the Royal Society of New Zealand for the Rutherford Discovery Fellowship that supported me while I wrote the manuscript.

Like many, I have always loved wandering along beaches, poking around in rock pools and picking up objects cast ashore by the waves. I have been lucky enough to build a career out of this passion, and hope that this book helps others to get to know a little more about the ocean and the life it holds. I think there's a marine biologist inside most people.

Index